Pulling Down Strongholds

A CRITIQUE OF THEISTIC EVOLUTION

Eric Parker
& Todd Hamilton, PhD

ONE STONE
BIBLICAL RESOURCES

Design, informational graphics and illustrations
by Stephen Sebree / Moonlight Graphic Works

Published by One Stone Press
979 Lovers Lane
Bowling Green, KY 42103
Printed in the United States of America

ISBN: 978-1-941422-92-2

ONE STONE
BIBLICAL RESOURCES

Dedication

I gratefully dedicate this book to those who instilled within my heart a love for preaching and furnished me with the skills to teach and preach the gospel of Jesus Christ. These influences include those professors who taught me at Florida College, too numerous to name, as well as those who sculpted me in the real world of preaching, most notably Alan Jones and Jim Jonas.

Ever in your debt,
EP

I dedicate this book to my beautiful wife, Lisa. You are the love of my life. And also our wonderful children, Ian and Olivia. We are so proud that you are letting your lights shine.

All my love,
TH

Foreword

The theory of evolution has become ingrained in the textbooks, museums, media and visitor centers of our National Parks. As these books and placards spread widely throughout the country in the 1990s and 2000s, new technology in the fields of molecular and cellular biology began to unlock countless cellular signaling pathways and complex mechanisms regulating gene expression. Such discoveries demonstrate complexity far beyond what Darwin could have ever imagined. The failure and/or resistance of the scientific community to reevaluate Darwin's theory based upon today's understanding of cellular biology is completely contrary to the way science should work. It is as if society desperately does not want to be under the authority of our Creator and as a result, they dig in their heels and refuse to accept any explanation other than evolutionary theory.

There comes a time in middle school, high school, or college when students may try to merge what they hear at school (evolution) with what they are taught at home (Creation by God/Intelligent Design). As parents, we raise our children to respect authority, but when they are taught evolution by authority figures such as teachers, students are sometimes conflicted. The idea of theistic evolution seems like a compromise between the two sides. As a young biology major in the 1990s, I recall hearing about theistic evolution, and initially it seemed reasonable. However, the more I learned about blood cells and the immune system in graduate school, it became very clear that there was no way these systems could have occurred by gradual changes over time. Resources such as this study book give parents, grandparents and teachers a tool to be prepared for discussions that arise as young people begin to question the origins of life.

Kristin M. Hall, PhD Microbiology and Immunology

Contents

CHAPTER 1

Biblical Creationism

The question of origins flummoxes mankind. Much more than a purely academic and philosophical pursuit, the issue of origins has bearing on who we are as humankind, what value we have (if any), and our relationship with what we observe. Answering the riddle of existence is of utmost importance in not only satisfying a troubling curiosity but also in confirming or overturning the way we see both ourselves and others.

This book presents a comparative analysis of the two primary origins options facing humanity: creationism and macroevolution. We say creationism in specific reference to biblical creationism, namely, the worldview that contends that humanity and all other aspects of the observable (and unobservable) order are the product of the Supreme Intelligence, Singular Builder, and High Designer of the universe (cf. Hebrews 3:4). In contrast, the worldview of macroevolution is a way of stating that the observable order is the product of countless epochs of time in which purely natural forces have mindlessly led to what we now experience. While some seek to homogenize these two worldviews, this book contends that to do so is to overlook, ignore, and contradict these mutually exclusive worldviews and belittle the unique right of biblical creationism to actuality.

In this first chapter, let's unpack the precise nature of biblical creationism, exploring the claims of the Bible regarding the origin of the universe.

The Nature of Our Creator

The theological labels above may, at first blush, seem too generic to have direct bearing on the Bible. The terms are valid, but it may be better to use less mechanical ideas to discuss our Maker. That is perhaps the more accurate approach of the Christian Scriptures, although, as noted in Hebrews 3:4 above, God is the Builder of all things except Himself.

The favored approach of the Bible in describing God's relationship to creation is that of Creator. This tactic is how the Bible opens: *"In the beginning, God created the heavens and the earth"* (Genesis 1:1). Note that the Bible does not commence with an argument for the existence of God; God's existence is stated as an unassailable fact. The Bible elsewhere openly condemns the dismissal of God's existence as the business of fools (e.g., Psalm 14:1). To worship or exalt anything else is to supplant His unique role as Creator and High King (Romans 1:18-25; Revelation 19:10).

But how exactly did God create? Or does the "how" even matter? Deity created our cosmos from nothing, a designation Christian Apologists label creation *ex nihilo* (cf. Romans 4:17; Hebrews 11:3). The Creator's ability to do this reveals several facets of His identity, truths that form the structural basis for fundamental biblical theology. God possesses complete self-reliance (i.e., aseity; cf. Acts 17:24-30), exists beyond the confines of space and time (i.e., omnipresence and eternality; cf. Psalm 90:1-2; 139:7-10), and is presumably unlimited in His oversight, knowledge, and power (i.e., sovereignty, omniscience, and omnipotence; cf. Psalm 139:1-6, 13-16).

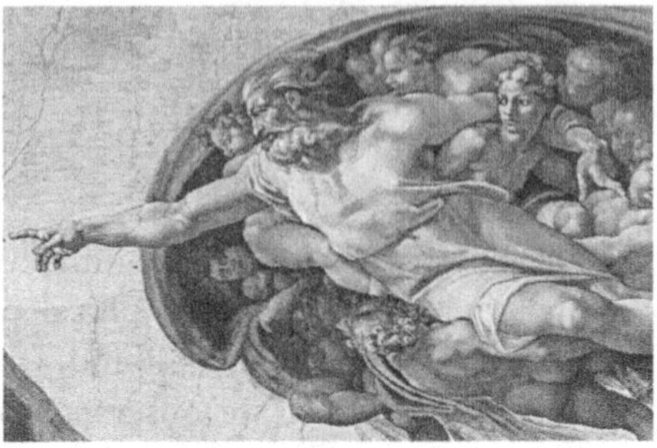

An outgrowth of God's identity as Creator is His role as Father. As the singular Progenitor of all else, God has given birth and life to all things (John 1:3).[1] But the role of father in this case is much more profound than God's role as the source of creation. His role as the God and Father of all mankind (Ephesians 4:6) means that He offers not only life to our current state of being but also an eternal home with Him in which we gain an immortal and incorruptible body in the resurrection (John 14:1-3; Romans 2:7, 10; 1 Corinthians 15:50-58).

The Biblical Nature of the Creation

Numerous passages in the Bible provide complementary details to the nature of creation. The most important and detailed passages include sections of Genesis, Job, and Psalms. Other miscellaneous passages augment the details in these passages but with less expansiveness.

Genesis 1-2

Genesis 1-2 offers God's opening presentation of creation, unpacking what exactly is meant by God's design as set forth in the very first verse of the Bible. Note the sweeping, yet incredibly brief, creation account of Genesis 1:2-2:4:

The earth was without form, and void; and darkness was on the face of

---

[1] Progenitor is an English term transliterated from Latin and indicates the source of one's existence.

the deep. And the Spirit of God was hovering over the face of the waters.

Then God said, "Let there be light"; and there was light. And God saw the light, that it was good; and God divided the light from the darkness. God called the light Day, and the darkness He called Night. So the evening and the morning were the first day.

Then God said, "Let there be a firmament in the midst of the waters, and let it divide the waters from the waters." Thus God made the firmament, and divided the waters which were under the firmament from the waters which were above the firmament; and it was so. And God called the firmament Heaven. So the evening and the morning were the second day.

Then God said, "Let the waters under the heavens be gathered together into one place, and let the dry land appear"; and it was so. And God called the dry land Earth, and the gathering together of the waters He called Seas. And God saw that it was good.

Then God said, "Let the earth bring forth grass, the herb that yields seed, and the fruit tree that yields fruit according to its kind, whose seed is in itself, on the earth"; and it was so. And the earth brought forth grass, the herb that yields seed according to its kind, and the tree that yields fruit, whose seed is in itself according to its kind. And God saw that it was good. So the evening and the morning were the third day.

Then God said, "Let there be lights in the firmament of the heavens to divide the day from the night; and let them be for signs and seasons, and for days and years; and let them be for lights in the firmament of the heavens to give light on the earth"; and it was so. Then God made two great lights: the greater light to rule the day, and the lesser light to rule the night. He made the stars also. God set them in the firmament of the heavens to give light on the earth, and to rule over the day and over the night, and to divide the light from the darkness. And God saw that it was good. So the evening and the morning were the fourth day.

Then God said, "Let the waters abound with an abundance of living creatures, and let birds fly above the earth across the face of the firmament of the heavens." So God created great sea creatures and every living thing that moves, with which the waters abounded, according to their kind, and every winged bird according to its kind. And God saw that it was good. And God blessed them, saying, "Be fruitful and multiply, and fill the waters in the seas, and let birds multiply on the earth." So the evening and the morning were the fifth day.

Then God said, "Let the earth bring forth the living creature according to its kind: cattle and creeping thing and beast of the earth, each according to its kind"; and it was so. And God made the beast of the earth according to its kind, cattle according to its kind, and everything that creeps on the earth according to its kind. And God saw that it was good.
Then God said, "Let Us make man in Our image, according to Our likeness; let them have dominion over the fish of the sea, over the birds of the air, and over the cattle, over all the earth and over every creeping thing that creeps on the earth." So God created man in His own image; in the image of God He created him; male and female He created them. Then God blessed them, and God said to them, "Be fruitful and multiply; fill the earth and subdue it; have dominion over the fish of the sea, over the birds of the air, and over every living thing that moves on the earth."

And God said, "See, I have given you every herb that yields seed which is on the face of all the earth, and every tree whose fruit yields seed; to you it shall be for food. Also, to every beast of the earth, to every bird of the air, and to everything that creeps on the earth, in which there is life, I have given every green herb for food"; and it was so. Then God saw everything that He had made, and indeed it was very good. So the evening and the morning were the sixth day.

Thus the heavens and the earth, and all the host of them, were finished. And on the seventh day God ended His work which He had done, and He rested on the seventh day from all His work which He had done. Then God blessed the seventh day and sanctified it, because in it He rested from all His work which God had created and made.

This is the history of the heavens and the earth when they were created, in the day that the Lord God made the earth and the heavens, before any plant of the field was in the earth and before any herb of the field had grown. For the Lord God had not caused it to rain on the earth, and there was no man to till the ground; but a mist went up from the earth and watered the whole face of the ground.

And the Lord God formed man of the dust of the ground, and breathed into his nostrils the breath of life; and man became a living being.

The structure of the creation account reveals to us the orderliness of God in contrast to the chaos of His absence. His production, organization, and purpose of creation is methodical, beautiful, and symmetrical. Readers will note the same rubric used for each day. Note the chart below, as well as a few subsequent observations regarding the text's bilateral symmetry:

Group I The Resource		**Group II The Utilizer**	
Day	*Creative Act*	*Day*	*Creative Act*
1	Light	4	The luminaries
2	Sky, leaving terrestrial waters	5	Fish and fowl
3	Dry land Vegtation (Lowest form of organic life)	6	Land creatures Humankind (Highest form of organic life)

- Opens and closes with the statement "God created";
- The third day is distinguished by two creative acts;
- Each set flows from the heavens to terrestrial water to dry land;
- Each act in the first set provides the resource for the corresponding acts in the second set.

This vital text of creationism shows God's creation of heavenly luminaries exceeding our comprehension down to the lowest forms of life. Readers also learn that humanity is given headship over creation, tasked with oversight and magnifying God's glory throughout His creation. This attribution and purpose of humanity—being imagers of God—subverts macroevolution, which delineates man as the end product, for now, of a long line of evolution devoid of purpose and meaning. There can be no starker contrast; there can be no harmony between the two viewpoints. Genesis 1-2 is clear – humankind is not the chance development of proteins devoid of existential value. Humankind is God's intentional, direct creation, invested with a magnificent position and destiny.

Job and Creation

The Book of Job is the definitive, in-depth treatment of the problem of suffering. It is such an impactful work that it is often studied outside of the ecclesiastical realm by those wrestling with the problem of evil. In conjunction with Genesis 3 (examined below), it is instrumental as a theodicy.[2] The book's first two chapters set Job's sufferings in a cosmic context, revealing that Job's issues were not limited to the seen world of flesh and blood. Job's woes were also a battle against foes and agents of darkness in the unseen realm (cf. Ephesians 6:10-17). Following Job 1-2, we see Job engage in a prolonged and heated dialogue with three friends – Eliphaz the Temanite, Bildad the Shuhite, and Zophar the Naamathite – about the nature of God and the problem of evil. All four parties, at times, make astute, wise remarks; at other times, they make asinine, rash, or overly simplistic statements. Following this heated debate, which runs from chapters 3-31, the perplexing figure Elihu enters the stage and offers his supplementary counsel across chapters 32-37. As Elihu rounded out his speech, he rolled out the red carpet for a lofty theophany in which God obliges Job in his challenge to show up and give a reason for his extensive suffering (chapters 38-42).[3] This theophany opens the door for God to offer Job a challenging admonition to grant God implicit trust based on Job's lack of perspective.[4]

In God's extensive rebuke of Job, He highlights a litany of items in the created order that Job has little to no knowledge of, let alone mastery over. Included in this list, we see God detail His creation, knowledge of, and dominion over the following:

- The seas and their boundaries (38:8-11);
- The sun, moon, and stars (38:12-15, 31-33);
- The realm of the dead (38:16-18);
- Ravens, lions, mountain goats, donkeys, wild oxen, horses, ostriches, birds of prey, etc. (38:39-39:30);
- Immensely powerful creatures like Behemoth and Leviathan (40:15-41:1ff).

The book of Job, and particularly this section of the book, shows that God's role as Master Builder is unimpeachable and also that it is foolish to elevate creatures over the Creator. As God Himself said in Job 38:4-7,

---

[2] A theodicy is a defense or justification (Greek: *dikē*) of God's (Greek: *theos*) benevolent nature despite the brokenness of our world.

[3] A theophany is an appearance (Greek: *phaneroō*) of God (Greek: *theos*). The Bible always depicts theophanies as a brief, humbling accommodation of God to impart knowledge, wisdom, and empowerment to His servants. These are not to be confused with the utterly unique incarnation of Jesus of Nazareth.

[4] All of this adds to earlier passing remarks regarding God's sovereignty and power (Job 26:7), which themselves illustrate scientific foreknowledge.

"Where were you when I laid the foundations of the earth?
Tell Me, if you have understanding.
Who determined its measurements?
Surely you know!
Or who stretched the line upon it?
To what were its foundations fastened?
Or who laid its cornerstone,
When the morning stars sang together,
And all the sons of God shouted for joy?"

Psalms and Creation

Psalms is a divinely-inspired collection of five individual books of songs used by the ancient Hebrews for various reasons and occasions.[5] Included among these was the opportunity to regularly reflect on God's majesty as experienced through His creation. Consider, for example, Psalm 8, which reads,

O Lord, our Lord,
How excellent is Your name in all the earth,
Who have set Your glory above the heavens!

Out of the mouth of babes and nursing infants
You have ordained strength,
Because of Your enemies,
That You may silence the enemy and the avenger.

[5] The five specific books are sometimes made evident in modern Bibles, but sometimes they are not. The books are divided as follows: Book I (Psalms 1-41), Book II (Psalms 42-72), Book III (Psalms 73-89), Book IV (Psalms 90-106), and Book V (Psalms 107-150).

When I consider Your heavens, the work of Your fingers,
The moon and the stars, which You have ordained,
What is man that You are mindful of him,
And the son of man that You visit him?
For You have made him a little lower than the angels,
And You have crowned him with glory and honor.

You have made him to have dominion over the works of Your hands;
You have put all things under his feet,
All sheep and oxen—
Even the beasts of the field,
The birds of the air,
And the fish of the sea
That pass through the paths of the seas.

O Lord, our Lord,
How excellent is Your name in all the earth!

Another reflection occurs in the first part of Psalm 19, which lays out the responsibility of humankind to their Creator as ascertained by general creation. As a counterpart, the second part of the psalm explains humanity's responsibility in light of special revelation in the Scriptures. Students of the Bible will note the shift from the general term for God in Hebrew (*El/Elohim*) in verses 1-6 to the use of the covenant name for God *(YHWH)* in verses 7-14. Note David's reflections in verses 1-6:

The heavens declare the glory of God;
And the firmament shows His handiwork.
Day unto day utters speech,
And night unto night reveals knowledge.
There is no speech nor language
Where their voice is not heard.
Their line has gone out through all the earth,
And their words to the end of the world.

In them He has set a tabernacle for the sun,
Which is like a bridegroom coming out of his chamber,
And rejoices like a strong man to run its race.
Its rising is from one end of heaven,
And its circuit to the other end;
And there is nothing hidden from its heat.

In addition to these personal reflections, psalmists reaffirm particular points of the Genesis narrative. Psalm 33:6-9 specifically praises God's voice as the agency of creation. It reads,

By the word of the Lord the heavens were made,
And all the host of them by the breath of His mouth.
He gathers the waters of the sea together as a heap;
He lays up the deep in storehouses.

Let all the earth fear the Lord;
Let all the inhabitants of the world stand in awe of Him.
For He spoke, and it was done;
He commanded, and it stood fast.

Asaph contemplates the oppression of his time in light of God's magisterial power and is confused by God's seeming inaction. Asaph called upon God for deliverance, meditating on God's creative power and blending cultural storylines and historical deliverances into his ruminations. Psalm 74:13-17 records Asaph's thoughts:

You divided the sea by Your strength;
You broke the heads of the sea serpents in the waters.
You broke the heads of Leviathan in pieces,
And gave him as food to the people inhabiting the wilderness.
You broke open the fountain and the flood;
You dried up mighty rivers.
The day is Yours, the night also is Yours;
You have prepared the light and the sun.
You have set all the borders of the earth;
You have made summer and winter.

At times, there are even lengthy psalms pondering God's role as Creator in which the psalmist offers specified poetic reflections of individual facets of creation. These considerations inevitably lead the psalmist to direct his own heart, as well as the hearts of those who sang the psalm, to worship the Lord in the raiment of humility and mindfulness. Psalm 104 is an excellent example of this. The Hebrews would sing this psalm and contemplate:

- God stretching out the heavens like a curtain, laying the beams of the upper chambers, and laying the foundations of the earth (verses 2-3, 5);

- God riding on the clouds as a chariot and walking on the wings of the wind, being served by angels ablaze with fire (verses 3-4);
- God withdrawing the sweeping floodwaters of the great deluge of Noah (verses 5-9);
- God daily providing abundant sustenance for mankind and beasts (verses 10-18);
- God's regulation of the time and the seasons (verses 19-23);
- God's command over all living things, both small and great, as His possessions (verses 24-30);
- God's worthiness to be praised for His righteous undertakings and power of fleshly and spiritual creation (verses 31-35).

The Corruption of God's Creation (Genesis 3)

Genesis not only offers instruction and explanation of the origin of the universe but also highlights significant changes that have interfered with the original created order. This explanation is especially helpful given the observable brokenness in our world. Gene decay, chromosomal aberrations, cancer, disease, and death present obvious challenges to the "good" state of creation, as repetitiously detailed in Genesis 1-2. Thankfully, the Genesis narrative proceeds to explain these and other significant degenerations from this wholesome state.

Genesis 3 details corruption's arrival into the creation picture. God's endowment of humanity with free will provided the opportunity for humanity to willingly choose to radiate the image of God by exercising delegated dominion over creation. Mankind's role of tending the Garden, exercising authority, and going forth, multiplying, and filling the earth constituted the positive expression of this liberty. God also communicated a prohibition: Adam and Eve were forbidden from partaking of the Tree of the Knowledge of Good and Evil (Genesis 2:16-17). God specified that it was not humanity's prerogative to partake of this tree; if they did, they would experience the death and decay that inevitably comes in transgressing God's will. Tempted by the Serpent, Eve partook of that fruit and gave to Adam, and he partook as well (cf. 1 Timothy 2:13-14). The three participants then began to experience the consequences of this departure. God explained that the world would experience a cataclysmic shift (Genesis 3:14ff):

So the Lord God said to the serpent:

"Because you have done this,
You are cursed more than all cattle,

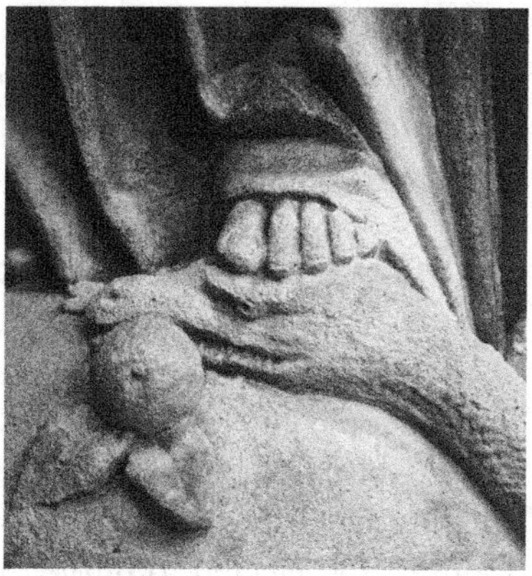

And more than every beast of the field;
On your belly you shall go,
And you shall eat dust
All the days of your life.
And I will put enmity
Between you and the woman,
And between your seed and her Seed;
He shall bruise your head,
And you shall bruise His heel."

To the woman He said:

"I will greatly multiply your sorrow and your conception;
In pain you shall bring forth children;
Your desire shall be for your husband,
And he shall rule over you."

Then to Adam He said, "Because you have heeded the voice of your wife, and have eaten from the tree of which I commanded you, saying, 'You shall not eat of it':

"Cursed is the ground for your sake;
In toil you shall eat of it
All the days of your life.
Both thorns and thistles it shall bring forth for you,
And you shall eat the herb of the field.
In the sweat of your face you shall eat bread
Till you return to the ground,
For out of it you were taken;
For dust you are,
And to dust you shall return."

And Adam called his wife's name Eve, because she was the mother of all living.

Also for Adam and his wife the Lord God made tunics of skin, and clothed them.

Then the Lord God said, "Behold, the man has become like one of Us, to know good and evil. And now, lest he put out his hand and take also of the tree of life, and eat, and live forever"— therefore the Lord God sent

> *him out of the garden of Eden to till the ground from which he was taken. So He drove out the man; and He placed cherubim at the east of the garden of Eden, and a flaming sword which turned every way, to guard the way to the tree of life.*

Genesis and the New Creation (Genesis 6-11)

As the world became more and more corrupt and degenerate, fewer and fewer people chose to observe God's will pursuing their own depravities instead. Genesis 6-9 details a world so captured by sin that God relented from creating it. His heartbreak is evident, especially in Genesis 6:1-7, 17. In response, God elected to purge the world of iniquity by means of a worldwide flood by which He could wholly purify the Earth. It was time for a new beginning.

However, the blameless man Noah and his household caught the eye of the Lord and offered an opportunity to restart through his line. God communicated the coming deluge, offering Noah plans to build a floating ark to serve as a vessel of grace and salvation (Genesis 6:8-10, 13-18). Noah and his sons Ham, Shem, and Japheth built the ark, preaching repentance to humanity. No one listened, and God sent the flood. In God's time, the fountains of the great deep erupted, and the windows of the heavens opened (see especially 7:11). All that was outside of the ark was destroyed and then reshaped into a new world. Disembarking from the ark after nearly a year aboard, Noah's family experienced the first rainbow, a symbol of God's promise not to exercise global devastation in the same way ever again. Noah's family and the animal kind representatives preserved within the ark spawned a new creation, and a new hope emerged that humanity would choose life.

As Noah's family and the animals repopulated the earth, mankind again chose to rebel (Genesis 10-11). Noah's descendants gathered in the plain of Shinar to exercise headship over the earth in their own way and after their own pattern. They began to build a tower to dethrone God and to take their seats as champions of the universe (Genesis 11:1-4). Again heartbroken, God split the peoples by dividing their language, driving them to the distant corners of the earth to fulfill His dominion mandate. The division of the nations then set the stage for God's election of Israel as His own special inheritance. Thus began the overt stages of the Messianic plan. Eve's seed, through Abraham's lineage, would bring life and restoration, a far greater new creation than anything experienced after the flood.

Concluding Thoughts

This chapter has surveyed the basic paradigm for biblical creationism. This is how the created order is described in the Scriptures, centuries before modern science and the general theory of evolution were advanced. Soon, we will compare biblical creationism with the general theory of evolution and explain why theistic evolution (i.e., believing in a created universe that then gradually evolved forward) is untenable. Before we do that, let's focus on how nature reveals the Creator and why believing otherwise is a perilous, inexcusable folly.

Thought/Discussion Questions

1. Is the biblical portrait of creation exclusive to evolution? Explain.

2. Where are the most extensive discussions of creation found in the Bible?

3. What does the Bible tell us about the Creator?

4. Do we need to look at natural phenomena and science to enhance our knowledge of the Creator?

5. Are there any discernible patterns in how God created life?

6. Why do you think the Bible neglects to provide an exhaustive, detailed description of every aspect of creation?

7. In what ways did the Flood of Noah point to a new creation?

CHAPTER 2

"They Are Without Excuse"

The mutually exclusive worldviews of creationism (aka supernaturalism) and atheism (aka naturalism) are laden with serious consequences. This is no trifling matter of relativism, opinion, and indifference. It is not a ground for compromise either. One option believes in a Creator God, and the other believes that all that exists is the mere product of nature. Creator or creation? Those are the options before each and every one of us when it comes to origins.

In this chapter, let's familiarize ourselves with the biblical condemnation of those who dismiss the reality of the Creator, choosing instead to deify nature and its processes. In particular, let's examine how the rise of scientism has led to widespread methodological naturalism and the a *priori* dismissal of God by those feigning objectivity.[6]

God Has Spoken: Who Will Listen?

As we noted in the previous chapter, God has not spoken in a distant corner where nobody could reasonably perceive His existence and relationship to the universe. Knowledge of God is made evident by what human beings can access and perceive through scientific inquiry, whether formal or informal. This means that God is personal and not some distant, impersonal, mystical, mysterious force. As one writer tacitly explains, God is not *"a nameless spirit awaiting post-mortem examination in some theological morgue. He is a very particular and specific divinity, known from the beginning solely on the basis of his works and self-declaration as the one living God."*[7] But how exactly is He known? What exactly is His chosen modality of self-declaration?

General Revelation

The first avenue of divine revelation is open for all to experience, even the illiterate. God has embedded His signature in the observable elements of our

---

[6]Methodological naturalism refers to restricting one's study and interpretation of the world to merely naturalistic processes. This common approach in science and philosophy arbitrarily disallows any connection to the divine.

[7]Carl F.H. Henry, God, *Revelation and Authority,* vol. 2: *God Who Speaks and Shows* (Waco: TX: Word, 1976), 7.

material universe. As human beings live in what ranges from the monotony of everyday existence to the once-in-a-million experiences that life can occasionally afford, God's existence is made knowable. Whether we stand at the heights of Everest, float in the Dead Sea, plumb the depths of the Mariana Trench, search for water in Death Valley, or gaze upwards at the great Redwoods in California, we can personally experience the vastness of God. Believers have accommodatively used the term "general revelation" to label such signatures in creation.

One example of this testimony is seen when we gaze at a newborn baby. Our thoughts and spirits are drawn together in intimate proximity. We contemplate the sanctity of human life. We consider the utter reliance of humanity on something more as we see this helpless infant in need of tender love and care and witness its parents provide for its basic needs. Who did that for the baby produced by macroevolution? Who was there for that child? God reminded His prophet Jeremiah of the One Who knew him before he was formed in the womb (Jeremiah 1:5). The psalmist also contemplated this aspect of God's handiwork (Psalm 139).

We might also look to the plant life God says He created on day three (Genesis 1:11-13). Whether we are gazing at the panicles of Teak trees in Myanmar, wondering at the immense bottle-shapes of Baobab trees in Kenya, or extracting the sustaining juices of Barrel cacti in the Sonoran Desert, we can see the skill of our God and King. Jesus even specially highlighted God's meticulous propagation and care of lilies of the field in the Sermon on the Mount. He connects God's care for such minutiae to His even more incredible care for humanity, warning about the folly of worry by saying,

> *"So why do you worry about clothing? Consider the lilies of the field, how they grow: they neither toil nor spin; and yet I say to you that even Solomon in all his glory was not arrayed like one of these. Now if God so clothes the grass of the field, which today is, and tomorrow is thrown into the oven, will He not much more clothe you, O you of little faith?"* (Matthew 6:28-30)

The righteous sufferer Job reproved his friends for their patronizing responses to his jaw-dropping afflictions, citing the fact that even plants and animals knew the truths these "paragons of wisdom" were offering. He appealed to this witness by saying:

> *"But now ask the beasts, and they will teach you;*
> *And the birds of the air, and they will tell you;*
> *Or speak to the earth, and it will teach you;*

And the fish of the sea will explain to you.
Who among all these does not know
That the hand of the Lord has done this,
In whose hand is the life of every living thing,
And the breath of all mankind?" (Job 12:7-10)

These are, of course, smaller illustrations in size. Consider the imposing items such as the sun, moon, stars, nebulae, and galaxies. Psalm 19 speaks of these wonders, directing the observer's attention to the One Who formed such phenomena. Some of the most fascinating signatures of God's design are evident in celestial contemplation. A pristine specimen of this is the fine-tuning that results in Earth being able to experience incredible eclipses. The proximity of the Earth to the Moon and the Sun allows us to be spectators of these infrequent and unique reminders of God's design. Note the detailed chart below, which illustrates several variables of this special distance relationship in contrast to other planets which do not experience these same wonders.

Lunar Eclipse Ratios for Our Solar System

	Planet-sun distance AU		Moon diameter km	Moon distance Mm	Diameter moon/sun
Earth	1	Moon	3476	384	1.0
Mars	1.52	Deimos	8	23	0.1
		Phobos	24	9	0.4
Jupiter	5.2	Europa	3126	671	2.6
		Io	3629	422	4.8
		Callisto	4800	188	14.3
		Ganymede	5276	1070	2.8
Saturn	9.54	Dione	1120	377	3.1
		Rhea	1528	527	3.0
		Titan	5150	1222	4.3
Uranus	19.19	Ariel	1160	191	12.5
		Oberon	1526	583	5.4
		Titania	1578	436	7.5
Neptune	30.07	Nereid	340	5513	0.2
		Proteus	420	118	11.6
		Triton	2705	355	24.7
Pluto	39.48	Charon	1207	20	262.5

The values of the ratio apparent diameter of moon/apparent diameter of the sun (from the point of view of an observer on the planet) range from 0.1 to 262.5. The earth/moon system is the only one with a value close to 1. AU = Astronomical Unit; the approximate distance between the earth and Sun. Mm = Megameters

On the other end of the spectrum, we enter the micro-world of bacteria, cells, and other infinitesimally small particles packed with incredible machinery that shames the finest examples of human engineering. From the flagellar motors of bacteria to the molecular machinery elemental to second-by-second processes that keep us alive, all of it points to the finger of God Almighty. The biochemist whose eyes and heart are open to see gains insight into the reality of the Great Biochemist.

Humanity has the great gift of senses by which they can look into the very face of God, observing His immutable power, His divine nature, and His invisible attributes (Romans 1:18-20). Even though these senses are detrimentally impacted by the sins plaguing this world, humans are still in a great position to use these senses employing science to gain insight into the Great Designer. As Augustine noted well, the Christian *"believes also the evidence of the senses which the mind uses by aid of the body; for if one who trusts his senses is sometimes deceived, he is more wretchedly deceived who fancies he should never trust them."*[8] In sum, God has spoken through general revelation; the question is, do we listen?

Special Revelation

Discerning God's impact on the physical world should lead us to seek an intimate relationship with Him. Our inquiry leads us to seek out special revelation. Humanity seeks a more direct, exhaustive, and relational communication from their Creator. We want answers regarding who we are in relation to Who made all of this. The inventor A.W. Dicus wrote a hymn about the intersection of general and special revelation, revealing their complementary role in our search for meaning. The four verses of his hymn and its refrain state:

There is, beyond the azure blue, a God concealed from human sight,
He tinted skies with heav'nly hue and framed the worlds with His great might.

There was, a long, long time ago, a God whose voice the prophets heard,
He is the God that we should know, who speaks from His inspired Word.

Secure is life from mortal mind, God holds the germ within His hand,
Tho' men may search, they cannot find, for God alone does understand.

---

[8]Augustine, *City of God*; trans. Marcus Dods (New York: Modern Library, 1950), 19.18.

Our God, whose Son upon the tree, a life was willing there to give, That He from sin might set man free, and evermore with Him could live.

There is a God, He is alive, in Him we live, and we survive; From dust our God created man, He is our God, the great I AM.

In several places in Scripture, God explains the obvious nature of His special revelation and contrasts it with the embarrassment of idolatry. Perhaps the most memorable of these treatments is found in Isaiah 45:18-25 which reads,

For thus says the Lord,
Who created the heavens,
Who is God,
Who formed the earth and made it,
Who has established it,
Who did not create it in vain,
Who formed it to be inhabited:
"I am the Lord, and there is no other.
I have not spoken in secret,
In a dark place of the earth;
I did not say to the seed of Jacob,
'Seek Me in vain';
I, the Lord, speak righteousness,
I declare things that are right.

Monarch butterflies migrating

"Assemble yourselves and come;
Draw near together,
You who have escaped from the nations.
They have no knowledge,
Who carry the wood of their carved image,
And pray to a god that cannot save.
Tell and bring forth your case;
Yes, let them take counsel together.
Who has declared this from ancient time?
Who has told it from that time?
Have not I, the Lord?
And there is no other God besides Me,
A just God and a Savior;
There is none besides Me.

> *"Look to Me, and be saved,*
> *All you ends of the earth!*
> *For I am God, and there is no other.*
> *I have sworn by Myself;*
> *The word has gone out of My mouth in righteousness,*
> *And shall not return,*
> *That to Me every knee shall bow,*
> *Every tongue shall take an oath.*
> *He shall say,*
> *'Surely in the Lord I have righteousness and strength.*
> *To Him men shall come,*
> *And all shall be ashamed*
> *Who are incensed against Him.*
> *In the Lord all the descendants of Israel*
> *Shall be justified, and shall glory.' "*

To our great benefit, God offers His divinely-inspired Scriptures. Creation points to our Creator God, and Scripture reveals explicit truths about God the Father, Jesus the Son, the Holy Spirit, creation, salvation, end times, etc. As the Apostle Paul wrote to Timothy: *"All Scripture is given by inspiration of God, and is profitable for doctrine, for reproof, for correction, for instruction in righteousness, that the man of God may be complete, thoroughly equipped for every good work"* (2 Timothy 3:16-17). In this revelation, God adequately and sufficiently equips us for our purpose. This resource manual leads us to a proper reverence for God (Ecclesiastes 12:13) and enables us to *"contend for the faith once for all delivered unto the saints"* (Jude 3).

Atheism: Choosing Not to Listen

Despite the open communication of God in revealing Himself to mortals, some choose not to believe in Him and formulate a view of the world devoid of God. This belief system is commonly known as atheism. William H. Halverson explains the stark contrast of atheism (aka naturalism) with theism in this way:

> *"Naturalism asserts, first of all, that the primary constituents of reality are material entities. By this I do not mean that only material entities exist; I am not denying the reality—the real existence—of such things as hopes, plans, behavior, language, logical inferences, and so on. What I am asserting, however, is that anything that is real is, in the last analysis, explicable as a material entity or as a form or function or action of a material entity.*

Theism says, 'In the beginning, God'; naturalism says, 'In the beginning, matter.' If the theoretical goal of science—an absolutely exhaustive knowledge of the natural world—were to be achieved, there would remain no reality of any other kind about which we might still be ignorant. The 'ultimate realities,' according to naturalism, are not the alleged objects of the inquiries of theologians; they are the entities that are the objects of investigations by chemists, physicists, and other scientists. To put the matter very simply: materialism is true.

Naturalism asserts, second, that what happens in the world is theoretically explicable without residue in terms of the internal structures and the external relations of these material entities. The world is, to use a very inadequate metaphor, like a gigantic machine whose parts are so numerous and whose processes are so complex that we have thus far been able to achieve only a very partial and fragmentary understanding of how it works. In principle, however, everything that occurs is ultimately explicable in terms of the properties and relations of the particles of which matter is composed. Once again the point may be stated simply: determinism is true."[9]

While some atheist groups passionately deny that atheism is a belief system or religion, it certainly fits the standard criteria, especially when we consider Halverson's remarks above.[10] Note this incisive critique of that dishonest claim:

"Almost all atheists claim that, because (supposedly) there is no God, their own worldview is not a religion. Many of them would argue that they have a 'nonbelief.'

One of the definitions of religion in the Merriam-Webster Dictionary, however, is this: 'a cause, principle, or system of beliefs held to with ardor and faith.'

Atheism certainly fits that definition, and many of its adherents are quite zealous about their faith system.

Atheists have an active belief system with views concerning origins (that

---

[9]William H. Halverson, *A Concise Introduction to Philosophy*, 3rd ed. (New York: Random House, 1976), 394.

[10]For an example, see: "What is Atheism?" *American Atheists*, https://www.atheists.org/activism/resources/about-atheism/

> *the universe and life arose by natural processes); no life after death; the existence of God; how to behave while alive; and so much more. Honest atheists will admit their worldview is a faith. Atheism is a religion!"*[11]

Theists are not the only ones to challenge the common viewpoint that atheism is not a belief system. One of the more intellectually honest atheists of our time was cellular biologist George Klein, who candidly wrote, "*My attitude is not based on science, but rather on faith . . . . The absence of a Creator, the non-existence of God is my childhood faith, my adult belief, unshakable and holy.*"[12]

Willful rebels who degrade themselves by this conclusion are without excuse. They have chosen to not know the knowable, dismiss the evident, and dishonor the One Who is worthy of all honor, glory, and dominion. The Holy Spirit speaks of such persons thus:

> *For the wrath of God is revealed from heaven against all ungodliness and unrighteousness of men, who suppress the truth in unrighteousness, because what may be known of God is manifest in them, for God has shown it to them. For since the creation of the world His invisible attributes are clearly seen, being understood by the things that are made, even His eternal power and Godhead, so that they are without excuse, because, although they knew God, they did not glorify Him as God, nor were thankful, but became futile in their thoughts, and their foolish hearts were darkened. Professing to be wise, they became fools, and changed the glory of the incorruptible God into an image made like corruptible man—and birds and four-footed animals and creeping things.*
>
> *Therefore God also gave them up to uncleanness, in the lusts of their hearts, to dishonor their bodies among themselves, who exchanged the truth of God for the lie, and worshiped and served the creature rather than the Creator, who is blessed forever. Amen.*
>
> *For this reason God gave them up to vile passions. For even their women exchanged the natural use for what is against nature. Likewise also the men, leaving the natural use of the woman, burned in their lust for one another, men with men committing what is shameful, and receiving in*

---

[11]Ken Ham and Bodie Hodge, "Atheism Is Religion," *Answers in Genesis* (February 20, 2017), https://answersingenesis.org/world-religions/atheism/atheism-is-religion/

[12]George Klein, *The Atheist in the Holy City* (Cambridge, MA: MIT Press, 1990), 203.

themselves the penalty of their error which was due.

And even as they did not like to retain God in their knowledge, God gave them over to a debased mind, to do those things which are not fitting; being filled with all unrighteousness, sexual immorality, wickedness, covetousness, maliciousness; full of envy, murder, strife, deceit, evil-mindedness; they are whisperers, backbiters, haters of God, violent, proud, boasters, inventors of evil things, disobedient to parents, undiscerning, untrustworthy, unloving, unforgiving, unmerciful; who, knowing the righteous judgment of God, that those who practice such things are deserving of death, not only do the same but also approve of those who practice them." (Romans 1:18-32)

Noting the immense weight of this choice, as well as the true motives behind this rebellion, one writer offers this assessment:

"To reverence the impersonal creation instead of the personal God who created us is a perversion designed for escaping moral accountability to the Creator. God indicts those who worship the creation instead of its Creator (Rom 1:18-23); and warns of the corruption of morals and behavior which result."[13]

Despite the ardor of atheists in attempting to avoid accountability, the light of God's undeniable revelation exposes their tenets of disbelief as folly. Upon examination, those tenets which appeared so fortified crumble (cf. Proverbs 18:17).[14] Northern Ireland theologian Alister McGrath reveals his own journey on this path of enlightenment: *"Atheism, I began to realize, rested on a less-than-satisfactory evidential basis. The arguments that had once seemed bold, decisive, and conclusive increasingly turned out to be circular, tentative, and uncertain."* [15]

---

[13]Dave Hunt, "The Greening of the Cross," *The Berean Call* (July 1997), https://www.thebereancall.org/content/greening-cross

[14]Consider the confident assertion of avowed atheist Sam Harris, *"Atheism is not a philosophy; it is not even a view of the world; it is simply an admission of the obvious."* (Sam Harris, *Letter to a Christian Nation* (New York: Knopf, 2006), 51). Harris and other representatives of New Atheism, such as Richard Dawkins, Daniel Dennett, and Christopher Hitchens, often portray religion, and specifically Christianity, as intrinsically dangerous, poisonous, and evil. How ironic.

[15]Alister McGrath, "Breaking the Science-Atheism Bond," https://www.beliefnet.com/news/science-religion/2005/08/breaking-the-science-atheism-bond.aspx

The Rise of Scientism

To appease the spiritual dissonance that necessarily comes from rejecting the Creator, many have sought refuge in the halls of science, bringing their atheistic presuppositions into the field and polluting the discipline. Science, in turn, has become dominated by atheists, who, having taken captive the field, erect a golden calf for the world to hail as their god (cf. Exodus 32). The tide has shifted, and public opinion of religion has fallen through this stratagem. Faith and science are now totally irreconcilable in the minds of many. Alister McGrath summarizes this movement of atheism in taking science as a discipline:

> *"The New Atheism frequently makes a rhetorical appeal to the natural sciences as the sole basis of reliable truth—a view now widely known as 'scientism'—and the concomitant rejection of religious belief as evidence-free superstition. On this view, science demands no commitments of faith; in fact, its emphasis on truth is the enemy of faith. The critical role of fiduciary judgements in the scientific method is simply airbrushed out of the picture. Although this theme is rhetorically echoed in the works of [Sam] Harris and [Christopher] Hitchens, it finds its fullest discussion in the works of [Richard] Dawkins and [Daniel] Dennett."*[16]

As the gap grew wider between theism and science, atheists began to smuggle their metaphysics into science, systematically rewriting the often neutral and inconclusive scientific narrative so that it inclines to a rigorous atheist conclusion. These metaphysical presuppositions are regularly portrayed as scientifically verified facts, bypassing the foundational methods of science. McGrath concludes, *"some of what is presented as 'science' by the New Atheism isn't really science at all: it actually amounts to the subversive manipulation of science by grafting on a patina of metaphysical naturalism."* [17] In another place, McGrath remarks,

---

[16]Alister E. McGrath. *Surprised by Meaning: Science, Faith, and How We Make Sense of Things* (Westminster: John Knox Press, 2011), 33.

[17]Ibid, 37.

"...when properly and legitimately applied, the scientific method is religiously neutral— neither uniformly supportive nor uniformly critical of religious beliefs. Since the scientific method clearly does not entail atheism, those who wish to exploit the sciences as a weapon against religion have to supplement their case in certain ways to increase its plausibility. One way is by using aggressive rhetoric, heaping ridicule on those who challenge its obvious inaccuracies. Another way is to rewrite the narrative of science, incorporating atheistic presuppositions into the narrative of the scientific method."[18]

What bearing does this have on the message of this book? Simply stated, the leveraging of science by atheists has culminated in one of two faulted responses on the part of believers: 1) Ignoring the findings of science, thinking that all science has an ax to grind against the Bible and God, and 2) Accepting atheistic science as gospel and corrupting biblical doctrines to submit to the altar of atheistic science (e.g., the literal days of creation). This is especially the case with the rise in popularity of theistic evolution. The fear of being labeled an anti-science buffoon is a real motivating force behind how Christians now approach science:

"Christians have been ostracized, mischaracterized, and ridiculed for years as being un-informed, un-thinking, and unwitting. This is particularly true during the post-Enlightenment period. Modernism and Post-modernism have taken center stage, dethroning theistic thinking with atheistic intellectual bravado. Science, in particular, has been crudely and incorrectly re-defined in terms of exclusively natural processes, and a hedge against supernatural involvement has been erected in the spirit of Babel's Tower.

As a result, all intellectual discussions quickly devolve into worldview discussions, namely the alleged superiority and intellectual prowess of atheistic naturalism. Recognition that these discussions have left the original subject matter and entered this plane is totally absent on the part of many naturalists. Rather, under the shroud of 'science,' Christians have been regarded as bumbling morons who are intellectually inept, irrational, and indisposed to accept the truth of scientific reason."[19]

---

[18]Ibid, 35.

[19]Eric Parker, *Behold the Builder: Scientific Evidences for the Biblical God* (Bowling Green, KY: One Stone Press, 2019), 10.

The Purpose of Science and the Struggle for Objectivity

We see then two extremes and an effort to bridge the gap based on compromise. The current state of affairs reflects a long history in which people have struggled to make sense of the world. For many, science has offered methods of answering life's big questions. McGrath describes science in this way: "*The natural sciences try to identify patterns within the natural world and then to seek out the deeper structures which can account for these patterns.*" [20] Science promises a systematic and objective way of looking at the available data. This makes sense in one way because some scientists do strive for their studies to be accurate, thoroughly tested, double-blind, reproducible, and verifiable. One can even understand not automatically invoking the supernatural realm when looking for systematic and statistical order in a straightforward, methodical, and thorough scientific investigation. However, science was never meant to be a worldview. It has been hijacked by atheists to supposedly fill the void of a belief system, a holistic philosophy of a person, and the built-in need of a person to believe in a higher power, worship, and have a personal relationship with a supernatural being (cf. Ecclesiastes 3:11).

We can look to the annals of the history of science to see the discipline used to support both belief and disbelief. Scientists such as Robert Grosseteste and Francis Bacon pioneered the scientific method in full conviction that the Creator has embedded measurable regularity in the world. Likewise, science trailblazers such as Johannes Kepler, Blaise Pascal, Robert Boyle, and Isaac Newton all professed great faith in the Creator. On the other side, you have scientists like Stephen Hawking, Jameel Sadik Al-Khalili, Niels Bohr, and other great minds who have been ardent atheists despite their profound effect on science. Then, of course, we have unforgettable science-religion conflicts like those of Galileo with the Roman Catholic Church over heliocentrism. The eruption of conflicts reveals that good science requires consistent rigorous safeguards, acknowledging the need for interpretation. These challenges, in particular, make objectivity in the doing of science incredibly difficult. That is why we observe such a range of beliefs in science.[21] Ronald Nash summarizes the foundations behind this struggle:

---

[20] McGrath, *Surprised by Meaning*, 22.

[21] Even here, though, the atheist's charged rhetoric creates a revisionist approach, overlooking the positive role of religion in the advancement of science. See, for example, the published works of the late particle physicist Victor Stenger, in which his claims obviously run aground against the actual history of science. His analysis of the historical intersection of religion and atheism in science is a clear case-in-point; see: Victor J. Stenger, *God and the Folly of Faith: The Incompatibility of Science and Religion* (Amherst, NY: Prometheus Books, 2012.

"A number of Christian writers have attempted to draw attention to the fact that theoretical thought is often strongly affected by nontheoretical considerations. It is hard to ignore the personal dimension to the acceptance and evaluation of world-views...Many people demonstrate that they are often incapable of thinking clearly about their world-view. Most of us have met people or read the writings of people who appear so captive to some conceptual scheme that they seem incapable of giving a fair hearing to any argument or piece of evidence that appears to threaten their cherished system...Sometimes people have difficulty with competing claims and systems because of philosophical presuppositions. But often, it seems clear, people's theoretical judgments seem inordinately affected by nontheoretical factors...theoretical thinking is never as pure or autonomous as many would like to think...it does seem that some people who appear to reject Christianity on rational grounds are, in fact, acting under the influence of nonrational factors, that is, more ultimate commitments of their hearts. People should be encouraged to dig below the surface and uncover the basic philosophical and religious presuppositions that often appear to control their thinking."[22]

In addition to these issues, there is also the challenge of what science has to inescapably assume to even begin its endeavor. Several factors seem prerequisite, a truth communicated well by both Ronald Nash and Alister McGrath:

"Even scientists make important epistemological, metaphysical, and ethical assumptions. They assume, for example, that knowledge is possible and that sense experience is reliable (epistemology), that the universe is regular (metaphysics), and that scientists should be honest (ethics). Without these assumptions that scientists cannot justify within the limits of their methodology, scientific inquiry would soon collapse."[23]

"The scientific method, when legitimately applied, simply does not take us to the destinations envisaged by the New Atheism. Science raises questions that it cannot answer, forcing us to ask deeper questions about this world and our place in it than the rather shallow rhetoric of recent populist atheism suggests...Science finds itself dependent upon certain working beliefs which are not 'susceptible of proof,' which [Thomas] Huxley rightly terms 'acts of faith.' For obvious reasons, New Atheist writers

---

[22]Ronald H. Nash, *Faith & Reason: Searching for a Rational Faith* (Grand Rapids: Zondervan, 1988), 28-29.

[23]Ibid, 27.

have been reluctant to draw attention to Huxley's insight, which contrasts sharply with the dogmatic insistence of Richard Dawkins that science is a faith-free zone."[24]

Concluding Thoughts

God has spoken in manifold ways accessible to all human beings. He has provided all the necessary components of a robust and fulfilling worldview. Sadly, some have elected, of their own free will, to ignore or compromise His revelation (2 Timothy 2:15). This rejection of the King of the universe cannot proceed without dire consequences. Those who willfully repudiate His holy communication are without excuse and God hands them over to the error of their ways (2 Peter 3:15-16). They will receive the full weight of rejecting the Creator in their own persons. As Karl Barth so pithily phrased this truth: "*The statement that 'God is dead' comes from Nietzsche and has recently been trumpeted abroad by some German and American theologians. But the good Lord has not died of this; He who dwells in the heavens laughs at them.*"[25]

In the next chapter, we will turn our direction to the systematic study of God's intelligent design of the universe. We will begin to unpack the light of God's revelation by thinking more acutely about His intricate and complex designs on micro and macro levels. But before we do that, we plead with all reading to consider your relationship with the Creator and His Word. You have every reason to entrust your soul to the One Who wove and knit you together in the womb. He knows each thought in your head and each word upon your lips. As Augustine so memorably taught:

> *"Great are you, O Lord, and exceedingly worthy of praise; your power is immense, and your wisdom beyond reckoning. And so we men, who are a due part of your creation, long to praise you – we also carry our mortality about with us, carry the evidence of our sin and with it the proof that you thwart the proud. You arouse us so that praising you may bring us joy, because you have made us and drawn us to yourself, and our heart is restless until it rests in you."*[26]

---

[24]McGrath, *Surprised by Meaning*, 37, 39.

[25]Karl Barth and Geoffrey W. Bromiley, *Letters 1961-1968* (Grand Rapids: William B. Eerdmans Publishing Company, 1975), 243.

[26]Augustine, *Confessions* Book 1, Chapter 1.

Thought/Discussion Questions

1. List some of the ways the Bible identifies denials of divine design.

2. Is it wrong or unloving for Christians to call out denials of divine design as foolish?

3. What are the two modes of divine communication? How are they similar? How are they different?

4. What were some of Alister McGrath's thoughts on science and religion?

5. Do scientists struggle with objectivity?

6. What is "scientism"?

7. What Bible passages would you add to support the content of this chapter?

CHAPTER 3

Intelligent Design and Creationism

We noted in our last chapter that the Bible gives an unflattering portrait of those who willfully suppress the truth of the existence of God (Psalm 14; Romans 1). In stark contrast, the Scriptures give a pleasing portrait of those who take the time to meditate on their Creator and the products of His creation (Psalms 8, 19). In this chapter, we want to build on this point by briefly tracing the systematic study of God as Designer and then focusing on the modern presentation of natural theology within the intelligent design (ID) movement.

A Brief History of Design Science

Historically, many influential scientists have used their skills to advance human knowledge and inspire appreciation of the Creator. They have done this primarily by focusing on the regularities, as well as the exceptions, of nature. This systematized study has often intended to inspire informed praise of God; indeed, for countless scientists, it could not reasonably lead to any other response! (cf. Proverbs 1:7; Romans 11:33) Note these quotes from influential science pioneers:

> **Isaac Newton:** *"This most beautiful system of the sun, planets and comets, could only proceed from the counsel and dominion of an intelligent and powerful Being. This Being governs all things, not as the soul of the world, but as Lord over all."*[27]

> **Galileo Galilei:** *"I do not feel obliged to believe that the same God who has endowed us with senses, reason, and intellect has intended us to forego their use."* [28]

> **Albert Einstein:** *"I want to know how God created this world. I am not*

[27]Isaac Newton, *Newton's Principia: The Mathematical Principles of Natural Philosophy*, translated by Andrew Motte (New York: Daniel Adee, 1825), 501.

[28]Galileo Galilei, *Letter to the Grand Duchess Christina of Tuscany* (1615). This letter is available in the public domain.

interested in this or that phenomenon, in the spectrum of this or that element. I want to know God's thoughts, the rest are details." [29]

The efforts of believing scientists to plumb these realities and to share them with others originated in the disciplines of teleology and natural theology with people like William Paley and the efforts of believing scientists to plumb these realities and to share them with others originated in the disciplines of teleology and natural theology with people like William Paley and have progressed down to the present in what is now commonly referred to as intelligent design.[30]

What Is Intelligent Design (ID)?

The Discovery Institute – one of the chief pioneering institutions funding and driving the development of intelligent design – offers this definition:

"Intelligent design refers to a scientific research program as well as a community of scientists, philosophers and other scholars who seek evidence of design in nature. The theory of intelligent design holds that certain features of the universe and of living things are best explained by an intelligent cause, not an undirected process such as natural selection."[31]

While some ID proponents are not biblically theistic, many are and have published regularly on the relationship between design in the world and the contents of the Bible, using their talents to advocate for creationism (cf. Matthew 25:14-30; Ecclesiastes 7:25a). ID promoters have regularly challenged the mindless processes of Darwinian Evolution (cf. Ephesians 5:11-13). Key influencers in this movement include: Michael Denton, Michael Behe, Stephen

---

[29]E. Salaman, "A Talk With Einstein," *The Listener* 54 (1955), 370-371.

[30]William Paley's landmark work *Natural Theology or Evidences of the Existence and Attributes of the Deity*, written in 1802 toward the end of his life, continues to be foundational reading for design apologists today. While some have criticized Paley for applying his principles too broadly and without discretion, particularly toward the latter half of his volume, his primary argument has never been successfully challenged.

[31]The think tank's Center for Science and Culture supplements this with their mission statement: *"The mission of Discovery Institute's Center for Science and Culture is to advance the understanding that human beings and nature are the result of intelligent design rather than a blind and undirected process. We seek long-term scientific and cultural change through cutting-edge scientific research and scholarship; education and training of young leaders; communication to the general public; and advocacy of academic freedom and free speech for scientists, teachers, and students."* (https://www.discovery.org/)

Meyer, Melissa Cain Travis, William Dembski, Gary Parker, Nancy Pearcey, and Phillip Johnson.[32] These and other ID promoters range across all sub-disciplines of academia and have all identified objects and processes within their fields of study that demonstrate God's handiwork. Denton and Behe, both biochemists, focus on cellular structure and molecular machinery that illustrate design. Meyer, a philosopher of science, discusses an assortment of variables across biology, geology, and even fossils. Travis uses her cutting-edge research to highlight elements of science, biology, and even philosophy that point to the Designer. Dembski is a mathematician who focuses on probability calculations as applied to the debate between creationism and naturalism. Parker uses his widespread knowledge, training, and experience across biology, as well as his conversion out of evolutionary theory, to the furtherance of creationism. The late Johnson was a trained lawyer who applied his law knowledge to trailblaze the modern ID movement through his authorship of *Darwin on Trial*. His seminal efforts earned him the honorary nickname "the father of intelligent design." These individuals and others have academically paved the way for the modern application of natural theology arguments by creationists (Proverbs 18:15).

---

[32]Their primary works include: Michael Denton, *Evolution: Still A Theory in Crisis* (Seattle: Discovery Institute Press, 2016); Michael J. Behe, *Darwin's Black Box: The Biochemical Challenge to Evolution* (New York: Simon & Schuster, 1996); Michael J. Behe, *Darwin Devolves: The New Science About DNA That Challenges Evolution* (New York: HarperOne, 2019); Michael J. Behe, *A Mousetrap for Darwin* (Seattle: Discovery Institute Press, 2020); Stephen C. Meyer, *Signature in the Cell: DNA and the Evidence for Intelligent Design* (New York: HarperOne, 2009); Stephen C. Meyer, *Darwin's Doubt: The Explosive Origin of Animal Life and the Case for Intelligent Design* (New York: HarperOne, 2014); Stephen C. Meyer, *The Return of the God Hypothesis: Three Scientific Discoveries That Reveal the Mind Behind the Universe* (New York: HarperOne, 2023); Melissa Cain Travis, *Thinking God's Thoughts: Johannes Kepler and the Miracle of Cosmic Comprehensibility* (Moscow, ID: Roman Roads Press, 2022); Melissa Cain Travis, *Science and the Mind of the Maker: What the Conversation Between Faith and Science Reveals About God* (Eugene, OR: Harvest House Publishers, 2018); William A. Dembski, *The Design Inference: Eliminating Chance Through Small Probabilities* (New York: Cambridge University Press, 1998); William A. Dembski, *No Free Lunch: Why Specified Complexity Cannot Be Purchased without Intelligence* (Lanham, MD: Rowman & Littlefield, 2001); *Debating Design: From Darwin to DNA*, edited by William Dembski and Michael Ruse (New York: Cambridge University Press, 2004); William A. Dembski and Jonathan Wells, *The Design of Life: Discovering Signs of Intelligence in Biological Systems* (Dallas: The Foundation for Thought and Ethics, 2008); Gary Parker, *Creation Facts of Life: How Real Science Reveals the Hand of God* (Green Forest, AR: Master Books, 2006); Nancy Pearcey, *Total Truth: Liberating Christianity from Its Cultural Captivity* (Wheaton, IL: Crossway, 2008); and Phillip E. Johnson, *Darwin on Trial* (Washington, D.C.: Regnery Gateway, 1991).

Foundational ID Principles

The Signature of Design

The most fundamental principle underlying the argument of ID is what many call the "signature of design." Simply put, when we ponder the human body, a flower, a landscape on a beautiful day, the sky on a clear night – we manifestly see beauty, organization, patterns, and encoded information. All of these examples exhibit undeniable, inherent, and visible design. Although naturalists argue that all we see appeared by fortuitous accident as the result of a "big bang" billions of years ago, common sense observation patently rules out that explanation.

Whenever we see something clearly designed, manufactured, and produced in our technological age—such as a watch, an automobile, or a computer program—we can imagine the creativity and the hours of planning and innovation necessary to bring it into existence. We observe the same level of intricacy and detail in nature, except on a much grander scale. Consider materials brought together by chance and undirected processes; the result of such activities is chaos, even destruction. It is apparent to all that intelligence and order are requisite to the creation of something that operates in a specified way. The Bible identifies God as that Cosmic Harmonizer (Genesis 1; Hebrews 3:4). This was the fundamental argument of William Paley when he composed his Watchmaker Argument.[33] This is also the humble acknowledgment of Charles Darwin: "*The impossibility of conceiving that this grand and wondrous universe, with our conscious selves, arose through chance, seems to me the chief argument for the existence of God.*"[34]

The sermons of Acts leverage this design signature observation as a pivotal stepping stone in beginning a relationship with the one true God. Paul and Barnabas strove to correct the idolatry of the residents of Lystra by making this point: God has left His fingerprints on this world so that we would direct our

---

[33]Even devout atheist Richard Dawkins is forced to admit the beauty of Paley's argument in his own refutation attempt. He wrote, "*Paley drives his point home with beautiful and reverent descriptions of the dissected machinery of life, beginning with the human eye....Paley's argument is made with passionate sincerity and is informed by the best biological scholarship of his day...one thing I shall not do is belittle the wonder of the living 'watches' that so inspired Paley.*" (Richard Dawkins, *The Blind Watchmaker* (London: W.W. Norton, 1985), 5.)

[34]"Darwin's Religion," *The Narracoorte Herald* (December 15, 1882), 3. Compare too Darwin's famous remark on the eye: "*To suppose that the eye, with all its inimitable contrivances for adjusting the focus to different distances, for admitting different amounts of light, and for the correction of spherical and chromatic aberration, could have formed by natural selection, seems, I freely confess, absurd in the highest possible degree.*" (Charles Darwin, *The Origin of Species*, 150th-anniversary edition (Alachua, FL: Bridge-Logos Foundation, 2009), 150).

worship to Him and not to the created order. Acts 14:15-17 states,

> *"Men, why are you doing these things? We also are men with the same nature as you, and preach to you that you should turn from these useless things to the living God, who made the heaven, the earth, the sea, and all things that are in them, who in bygone generations allowed all nations to walk in their own ways. Nevertheless He did not leave Himself without witness, in that He did good, gave us rain from heaven and fruitful seasons, filling our hearts with food and gladness."*

Later, in Acts 17, Paul gained the opportunity to speak to the Areopagus in Athens. His message again concentrated on the fact that creation points conclusively to the one true Creator. In Acts 17:22-31, Paul took his stand and preached,

> *"Men of Athens, I perceive that in all things you are very religious; for as I was passing through and considering the objects of your worship, I even found an altar with this inscription:*
>
> *TO THE UNKNOWN GOD.*
>
> *Therefore, the One whom you worship without knowing, Him I proclaim to you: God, who made the world and everything in it, since He is Lord of heaven and earth, does not dwell in temples made with hands. Nor is He worshiped with men's hands, as though He needed anything, since He gives to all life, breath, and all things. And He has made from one blood every nation of men to dwell on all the face of the earth, and has determined their preappointed times and the boundaries of their dwellings, so that they should seek the Lord, in the hope that they might grope for Him and find Him, though He is not far from each one of us; for in Him we live and move and have our being, as also some of your own poets have said, 'For we are also His offspring.' Therefore, since we are the offspring of God, we ought not to think that the Divine Nature is like gold or silver or stone, something shaped by art and man's devising. Truly, these times of ignorance God overlooked, but now commands all men everywhere to repent, because He has appointed a day on which He will judge the world in righteousness by the Man whom He has ordained. He has given assurance of this to all by raising Him from the dead."*

Irreducible and Specified Complexity

This brings us to the next major component of the ID argument – irreducible and specified complexity. In 1996, Michael Behe authored his groundbreaking work *Darwin's Black Box: The Biochemical Challenge to Evolution.* His cornerstone argument was that science has discovered numerous systems that exhibit a Designer and that these systems defy the capabilities of evolutionary development.[35] In his book, he deftly presents a series of biological systems, comparing them to Rube Goldberg machines, driving the point home that the removal of parts in these systems would result in mechanical failure. Perhaps his most memorable illustration is in Chapter 2 of his work, where he analogizes with a traditional household mouse trap, in which the removal of the platform, hammer, spring, sensitive catch, or holding bar would render the mousetrap useless.[36] He then sums up the idea of irreducible complexity in this way:

> *"By irreducibly complex I mean a single system composed of several well-matched, interacting parts that contribute to the basic function, wherein the removal of any one of the parts causes the system to effectively cease functioning. An irreducibly complex system cannot be produced directly (that is, by continuously improving the initial function, which continues to work by the same mechanism) by slight, successive modifications of a precursor system, because any precursor to an irreducibly complex system that is missing a part is by definition nonfunctional. An irreducibly complex biological system, if there is such a thing, would be a powerful challenge to Darwinian evolution. Since natural selection can only choose systems that are already working, then if a biological system cannot be produced gradually it would have to arise as an integrated unit, in one fell swoop, for natural selection to have anything to act on."*[37]

---

[35]Behe is not the first scientist to describe such systems. Predecessors such as Gustav Cuvier wrote of the same observation 150 years prior: *"Every organized being forms a whole, a unique, and perfect system, the parts of which mutually correspond, and concur in the same definitive action by a reciprocal reaction. None of these parts can change without the whole changing; and consequently each of them, separately considered, points out and marks all the others."* (G. Cuvier, *A Discourse on the Revolutions of the Surface of the Globe and the Changes thereby Produced in the Animal Kingdom* (Philadelphia: Carey & Lea, 1831), 59.)

[36]Attempts to upend this illustration and other facets of Behe's argument regarding irreducible complexity have been given sufficient rebuttal in: Michael J. Behe, *A Mousetrap for Darwin* (Seattle: Discovery Institute Press, 2020).

[37]Behe, *Darwin's Black Box*, 39.

The million-dollar question then is how to detect, test, and measure Behe's notion of irreducible complexity so that we can evaluate the merits of any possible prospects for the claim. To put it differently, how do we go from the sweeping, undetailed observations that the Scriptures present as evidence of design (e.g., Job 38-41) to the meticulously detailed observations of a post-Enlightenment laboratory? Can we even do that? Behe offers his own thoughts:

> *"The scientific problem then becomes, how do we confidently detect design? When is it reasonable to conclude, in the absence of firsthand knowledge or eyewitness accounts, that something has been designed? For discrete physical systems—if there is not a gradual route to their production—design is evident when a number of separate, interacting components are ordered in such a way as to accomplish a function beyond the individual components. The greater the specificity of the interacting components required to produce the function, the greater is our confidence in the conclusion of design."*[38]

William Dembski is another prominent academician consistently publishing on the merits of the ID argument. He also refers to the combining of complexity with the specification of codependent components to accomplish their designated duty. He refers to this combination as the *"key feature of life that needs to be explained."*[39] He explains,

> *"Life is both complex and specified. The basic intuition here is straightforward. A single letter of the alphabet is specified without being complex (i.e., it conforms to an independently given pattern but is simple). A long sequence of random letters is complex without being specified (i.e., it requires a complicated instruction-set to characterize but conforms to no independently given pattern). A Shakespearean sonnet is both complex and specified."*[40]

---

[38]Ibid, 194. Compare William Dembski's similar analysis and conclusion: *"My project, then, was to formalize and extend our commonsense understanding of design inferences so that they could be rigorously applied in scientific investigation...At the heart of my codification of design inferences was the combination of two things: improbability and specification. Improbability, as we shall see in the next section, can be conceived as a form of complexity. As a consequence, the name for this combination of improbability and specification that has now stuck is specified complexity or complex specified information."* (William Dembski, "Explaining Specified Complexity," (September 13, 1999), https://metanexus.net/explaining-specified-complexity/).

[39]Ibid.

[40]Ibid.

Dembski then proceeds to statistically detail how the macroevolutionary paradigm falls abysmally short in explaining specified complexity. He determines that no evolutionary algorithm (e.g., the Darwinian mutation-selection mechanism, neural nets, and genetic algorithms) has been offered to explain specified complexity convincingly. He even evaluates Richard Dawkins' paradigmatic "Me Thinks It Is Like A Weasel" 43-step cumulative selection process, which Dawkins purports to unhinge the specified complexity argument.[41] Dembski concludes:

> *"It follows that Dawkins's evolutionary algorithm, by vastly increasing the probability of getting the target sequence, vastly decreases the complexity inherent in that sequence. As the sole possibility that Dawkins's evolutionary algorithm can attain, the target sequence in fact has minimal complexity (i.e., the probability is 1 and the complexity, as measured by the usual information measure, is 0). In general, then, evolutionary algorithms generate not true complexity but only the appearance of complexity. And since they cannot generate complexity, they cannot generate specified complexity either.*
>
> *There's only one known source for producing actual specified complexity, and that's intelligence. In every case where we know the causal history responsible for an instance of specified complexity, an intelligent agent was involved. Most human artifacts, from Shakespearean sonnets to Durer woodcuts to Cray supercomputers, are specified and complex. For a signal from outer space to convince astronomers that extraterrestrial life is real, it too will have to be complex and specified, thus indicating that the extraterrestrial is not only alive but also intelligent (hence the search for extraterrestrial intelligence-SETI)...In the case of Darwinism, what I have established is that the Darwinian mechanism cannot generate actual specified complexity."*[42]

Fine-Tuning

The *coup de grace* of ID arguments is evident and measurable fine-tuning. Two scientists from two different worldviews discuss this component:

---

[41]Richard Dawkins, *The Blind Watchmaker*, 47-48.

[42]Dembski, "Explaining Specified Complexity." For an in-depth analysis of the intelligence requirement for specified complexity, see: William A. Dembski, *No Free Lunch: Why Specified Complexity Cannot Be Purchased without Intelligence* (Lanham, MD: Rowman & Littlefield, 2007).

Paul Davies (an agnostic physicist): *"There is for me powerful evidence that there is something going on behind it all...It seems as though somebody has fine-tuned nature's numbers to make the Universe...The impression of design is overwhelming."* [43]

Ed Harrison (a Paleyan deist): *"Here is the cosmological proof of the existence of god - the design argument of Paley - updated and refurbished. The fine tuning of the universe provides prima facie evidence of deistic design. Take your choice: blind chance that requires multitudes of universes or design that requires only one....Many scientists, when they admit their views, incline toward the teleological or design argument."*[44]

When we look at the various components of our universe, we not only see that systems get the job done, but overwhelmingly, they get the job done well by means of regulatory networks like encoded information and mapping protocols (cf. Isaiah 45:18-19).[45] Some have employed this idea in the argument of design from anthropic constants; others have used the concept in arguing for the origin of biological systems; and still others have applied the observable principle across other fields of scientific inquiry.[46] Agnostics and atheists have also attempted to explain (or explain away) these haunting matters of fine-tuning.[47]

Let's illustrate with an example. Think of Fred Flintstone's foot-powered Flintmobile versus a Bugatti Veyron. Fred's vehicle will get him to places (at least in cartoon theory), but we can hardly say that his car is fine-tuned, even if it does get the basic job done. In contrast, the Veyron is a fine-tuned super-machine. The 8.0 Liter quad-turbocharged engine with 7-speed dual-clutch trans-

---

[43]Paul Davies, *The Cosmic Blueprint: New Discoveries in Nature's Creative Ability to Order the Universe* (West Conshohocken, PA: Templeton Foundation Press, 2004), 203.

[44]Edward Harrison, *Masks of the Universe: Changing Ideas on the Nature of the Cosmos* (New York: Cambridge University Press, 2003), 54, 252, 263.

[45]Exceptions, real or perceived, to this general truth do not undermine the argument. The fact that some things are broken or less than ideal is a philosophical matter. A Rolex watch found in the woods with a cracked screen or a dead battery still exemplifies design and begs an explanation. The problem of evil, brokenness, pain, and suffering is addressed elsewhere in detail, especially in the works of C.S. Lewis, Ronald Nash, and Philip Yancey.

[46]For a believer's example, see: Stephen C. Meyer, *Return of the God Hypothesis*.

[47]For an agnostic example, see: Paul Davies, *The Goldilocks Enigma: Why is the Universe Just Right for Life?* (Boston: Mariner Books, 2008). For an atheist example, see: Victor Stenger, *The Fallacy of Fine-Tuning: Why the Universe Is Not Designed For Us* (Amherst, NY: Prometheus Books, 2011).

mission packs a walloping 1,000+ horsepower, making the Stone-Age vehicle laughable. Engineers spent years formulating the exact specifications of the supercar to ensure it was top-of-the-line. That is the kind of universe that we live in—a universe which openly exhibits engineering, intention, and encoding.[48]

ID Icons (1 Kings 4:29-34)

Design scientists and writers have found countless biological systems that defy evolutionary explanations. This is true both on the macro and the micro level. Many of these designs have inspired engineers and advanced conceptual understanding in ways that have facilitated exponential technological advancement (e.g., biomimicry). Let's take a few moments to discuss some of the systems most common to design science literature.

Bacterial Flagellum (John 1:3)

Let's begin with what has quickly become the go-to ID illustration—the bacterial flagellum. The bacterial flagellum is a whiplike motility device found across many species of bacteria, both gram-positive and gram-negative. Approximately thirty interdependent parts comprise this minuscule propulsion system. The third chapter of Behe's *Darwin's Black Box* contains one of the first systematic examinations of the bacterial flagellum (along with bacterial cilia). He has also written extensively on the bacterial flagellum since that publication. In one source, he describes it this way,

> *"The flagellum can be thought of as an outboard motor that bacteria use to swim. It was the first truly rotary structure discovered in nature. It consists of a long filamentous tail that acts as a propeller; when it is spun it pushes against the liquid medium and can propel the bacterium forward. The propeller is attached to the drive shaft indirectly through something called the hook region, which acts as a universal joint. The drive shaft is attached to the motor, which uses a flow of acid or sodium ions from the outside of the cell to the inside to power rotation. Just as an outboard motor has to be kept stationary on a motorboat while the propeller turns, there are proteins which act as a stator structure to keep the flagellum in place. Other proteins act as bushings to permit the drive shaft to pass through the bacterial membrane. Studies have shown that 30-40 proteins are required to produce a functioning flagellum in the cell.*

---

[48]For more information on the role of information in the argument of ID, see: Werner Gitt, *In the Beginning Was Information: A Scientist Explains the Incredible Design in Nature* (Green Forest, AR: Master Books, 2006).

About half of the proteins are components of the finished structure, while the others are necessary for the construction of the flagellum. In the absence of almost any of the proteins—in the absence of the parts that act as the propeller, drive shaft, hook, and so forth—no functioning flagellum is built."[49]

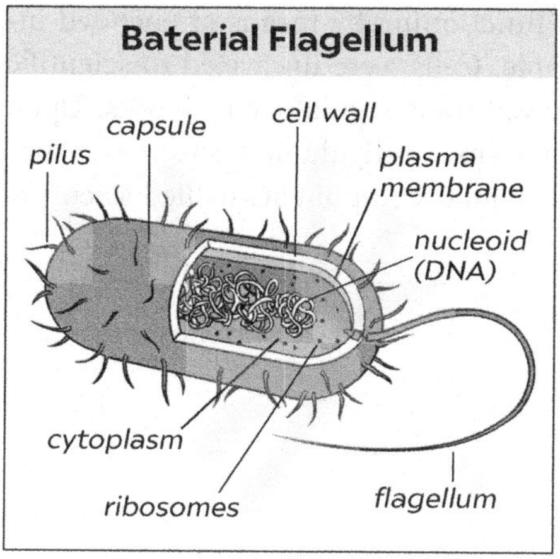

This ID paragon is not without its critics. Behe's interlocutors have suggested that the bacterial flagellum is not an irreducibly complex system, as he contends, but that the simpler, ten-component Type III Secretory System is the evolutionary pathway by which the bacterial flagellum developed. Although this response has become popularly accepted, it is riddled with insurmountable issues.[50] Mathematician William Dembski offers this conclusion on the matter:

"Is the claim that the bacterial flagellum exhibits specified complexity assertible? You bet! Science works on the basis of available evidence, not on the promise or possibility of future evidence. Our best evidence points to the specified complexity (and therefore design) of the bacterial flagellum. It is therefore incumbent on the scientific community to admit,

---

[49]Michael J. Behe, "Irreducible Complexity: Obstacle to Darwinian Evolution," in *Debating Design: From Darwin to DNA*, edited by William A. Dembski and Michael Ruse (New York: Cambridge University Press, 2004), 353.

[50]Studies concerning mutation density studies, phylogenetic distribution, plasmid problems, and the age of eukaryotic cells show that the bacterial flagellum is older than the Type III secretory system and that the latter is either a degenerative byproduct of the bacterial flagellum or else independent. For a more detailed explanation, see: "Irreducible Complexity, Bacterial Flagellum, and the Type III Secretory System," *Discovery Science* (June 2, 2020), https://www.youtube.com/watch?v=G581HlqXSFg

at least provisionally, that the bacterial flagellum could be the product of design."[51]

Cells, DNA, and RNA (Colossians 1:16)

Advances in microscopy in the early-to-mid-twentieth century opened up a world of incredible micro-machines functioning by means of encoded information formerly assumed to be simple. Cells were unraveled to scientific inquiry, and molecular machinery showed their regulatory networks. Upon close inspection, mitochondria, endoplasmic reticulum, lysosomes, ribosomes, nuclei, and other incredible subcellular components baffled scientists with their awe-inspiring composition.[52]

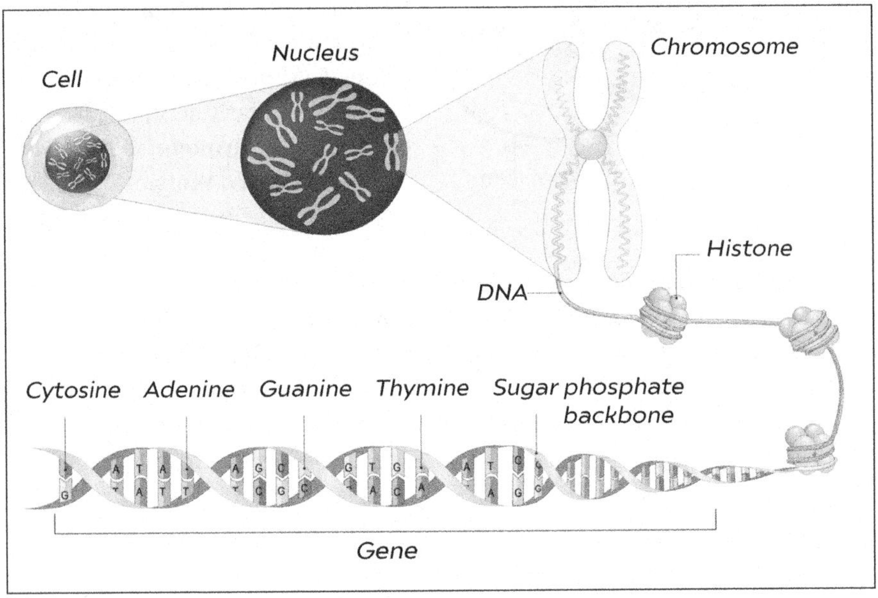

As Watson and Crick unlocked the double helix structure of deoxyribonucleic acid (DNA) and scientists began understanding the nitrogen bases adenine, thymine, cytosine, and guanine, the struggle for naturalistic explanations of biological origins became more and more pronounced. DNA's quaternary system, the backbone of natural life, made the binary system of computers

[51]William A. Dembski, "The Logical Underpinnings of Intelligent Design," in *Debating Design: From Darwin to DNA*, edited by William A. Dembski and Michael Ruse (New York: Cambridge University Press, 2004), 326.

[52]For an assessment of cellular design, see Fazale Rana, *The Cell's Design: How Chemistry Reveals the Creator's Artistry* (Baker Books, 2008). See also: Meyer, *Signature in the Cell*.

written by the most intelligent minds in the world look like amateur hour. As scientists performed more research on DNA, the supportive functions of RNA and epigenetics also became more evident. The enigma of the microscopic world captured the attention of scientists as natural theology once more became rationally incontestable. One source speaks of the developing dilemma facing God-deniers this way:

> *"The incredible specified complexity of life becomes obvious when one considers the message found in the DNA of a one-celled amoeba (a creature so small, several hundred could be lined up in an inch). Staunch Darwinist Richard Dawkins, professor of zoology at Oxford University, admits that the message found in just the cell nucleus of a tiny amoeba is more than all thirty volumes of the Encyclopedia Britannica combined, and the entire amoeba has as much information in its DNA as 1,000 complete sets of the Encyclopedia Britannica! In other words, if you were to spell out all of the A, T, C, and G in the 'unjustly called 'primitive' amoeba' (as Dawkins describes it), the letters would fill 1,000 complete sets of an encyclopedia!*
>
> *Now, we must emphasize that these 1,000 encyclopedias do not consist of random letters but of letters in a very specific order—just like real encyclopedias. So here's the key question for Darwinists like Dawkins: if simple messages such as "Take out the garbage—Mom,' 'Mary loves Scott,' and 'Drink Coke' require an intelligent being, then why doesn't a message 1,000 encyclopedias long require one?"*[53]

Blood Clotting (Leviticus 17:11)

Another central model of ID is the blood clotting cascade, a biological system necessary to stop bleeding. The blood clotting cascade is a system that contains nearly two dozen inter-reliant factors in order to function correctly. In light of this complex mechanism, consider evolutionary theory, which contends that the blood clotting cascade is the product of the gradual combining of independent systems. How is it possible that evolution could provide a pathway forward without blood clotting already being in place? How could our biological antecedents survive long enough to reproduce without blood clotting already in place? One cut, one wound, and that's it. End of the road.

---

[53]Norman S. Geisler and Frank Turek, *I Don't Have Enough Faith to Be an Atheist* (Wheaton, IL: Crossway, 2004), 116. Geisler and Turek are quoting from: Richard Dawkins, *The Blind Watchmaker* (New York: Norton, 1987), 17-18, 116.

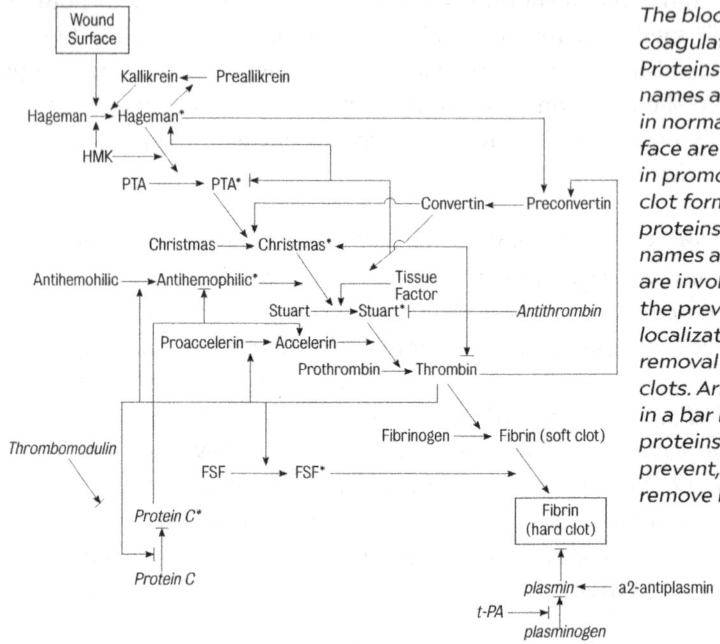

The blood coagulation cascade. Proteins whose names are shown in normal type face are involved in promoting clot formation; proteins whose names are italicized are involved in the prevention, localization, or removal of blood clots. Arrows ending in a bar indicate proteins acting to prevent, localize, or remove blood clots.

Behe summarizes the beauty of the blood clotting cascade this way:

> *"Like some ultimate Rube Goldberg machine, the clotting cascade is a breathtaking balancing act in which a menagerie of biochemicals—sporting various decorations and rearrangements conferred by modifying enzymes—bounce off one another at precise angles in a meticulously ordered sequence until, at the denouement, Foghorn Leghorn pushes off the telephone pole and gets up from the ground, the bleeding from his wounds stopped. The audience rises to its feet in sustained applause."*[54]

Woodpeckers, Bombardier Beetles, and Planthoppers (Genesis 1:20-25)

In addition to these intricate microscopic information systems, design advocates have drawn attention to ID exemplars among particular animal species.

The formerly outspoken evolutionist Gary Parker has highlighted the superiority of creationism over macroevolution in explaining woodpeckers. Speaking of irreducible complexity in biological organisms, he writes:

[54]Behe, *Darwin's Black Box*, 97.

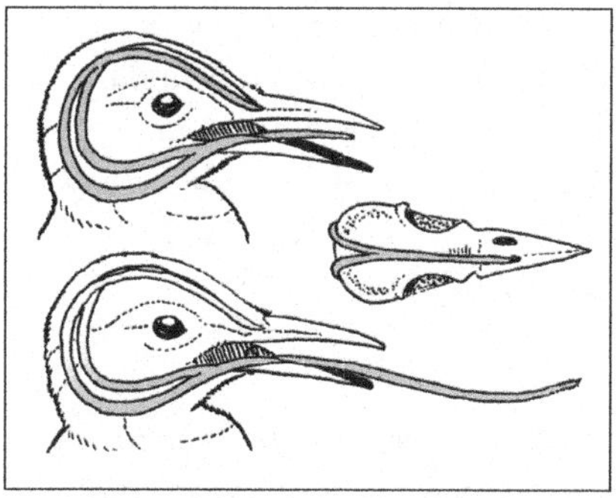

"Take the woodpecker for instance. Here's a bird that makes its living banging its head into trees...How did banging its head into trees increase its likelihood for survival – until after it had accumulated (by chance?) a thick skull with shock-absorbing tissues, muscles, etc.! ... What would be the survival value of all these features (and how could they build up in the population) until after the bird started banging its head into trees? The woodpecker is a marvel of interdependent parts or 'compound traits,' now popularly called 'irreducible complexity' – traits that depend on one another for any to have functional value...For such drilling, a woodpecker obviously needs a tough bill, heavy duty skull, and shock-absorbing tissue between the two...In fact, neither the tough bill nor the heavy-duty skull would have any functional survival value until both occurred together – along with the shock-absorbing tissue, nerve and muscle coordination, etc.! That's no problem if the woodpecker were put together by plan, purpose, and a special act of creation...There's more...some woodpeckers (are) looking for bark beetles. The beetles hear all this pounding, of course, so they just crawl further down their tunnels. Some...woodpeckers need more than just drilling tools; they need long, sticky tongues. But...what's it going to do with it? The answer for the woodpecker is to slip its tongue attachment into a muscular sheath that wraps around the skull under the scalp and inserts into the nostril! That makes good sense (and good science) if you're planning ahead, but poses real problems if your faith is in time and chance, trial and error."[55]

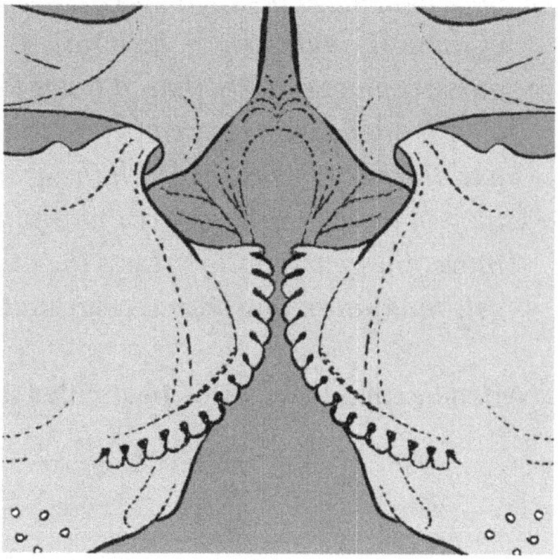

Planthopper's gears

Speaking of beetles, researchers have also highlighted the chemical expulsion systems utilized by bombardier beetles as illustrative of design. Gary Parker writes,

[55]Parker, *Creation Facts of Life*, 63, 65-66.

"The bombardier (beetle) is an ordinary looking beetle, but it has an ingenious chemical defense mechanism. Just as (a toad) gets ready to flash out that long, sticky tongue, the beetle swings its cannon around, and 'boom!' It blasts the toad in the face with hot noxious gases at the boiling point of water, and coats the toad's tongue with a foul-tasting residue... Pictures show the toad dragging its tongue across the sand trying to get rid of the foul taste.

Successful firing of the bombardier beetle's cannon requires two chemicals (hydrogen peroxide and hydroquinones), enzymes, pressure tanks, and a whole series of nerve and muscle attachments for aim and control. Try to imagine all those parts accumulating by time, chance, and natural selection. One crucial mistake, of course, and 'boom!'...the beetle blows itself up, and there's no evolutionary future in that!"[56]

Among the bug world, scientists have also discovered a fascinating synchronization gear system in adolescent planthoppers. One biochemist describes these amazing engineering components:

"A recent study discovered a phenomenal secret of one of the juvenile stages of the species Issus coeleoptratus. It had been reported a half century ago that strange bumps occurred on the hind legs of young planthoppers, but no purpose had ever been assigned to them...A pair of British entomologists, armed with sophisticated high-speed video equipment, showed that the bumps are actually the teeth of gears...For the planthopper to achieve the high-speed takeoff velocity needed to jump hundreds of times its body length, its hind legs must begin to flex in synchrony very quickly, more quickly than it takes for a full nerve impulse to reach the legs. If one leg is triggered before the other, the insect would lose power and tumble erratically. With the gear teeth engaged and the gears spinning at an astonishing fifty thousand teeth per second, as one leg starts to move, the gear rotation starts the other leg moving as well, and the bug gets maximum power and coordination for its efforts." [57]

Behe explains why this particular structure has been so impactful:

---

[56]Ibid, 100.

[57]Behe, *Darwin Devolves*, 45.

> *"Although mechanical devices are a dime a dozen at the molecular and cellular levels of life, the planthopper's equipment is the first example of a (relatively) large, in-your-face, interacting gear system in an animal. In an interview with National Geographic, one of the authors of the study was betting it wouldn't be the last: 'There is stuff that's vastly more intricate and complicated that hasn't been found yet.' Despite the immense recent progress of biology, he is surely right. There seems to be a bottomless supply of wonders in life. Expect plenty more where that came from!*
>
> *The* National Geographic *writer was thrilled: 'This insect has gears! GEARS!' That sound you hear is the jaws of thousands of readers hitting the floor. But why the astonishment? There are plenty of more sophisticated and admirable organs in nature, including the brain—the organ people use to help grasp the significance of all the others. So why is the discovery of gears so sensational? I think one of the reasons is that the purpose is so plain to see. Despite the sophistication of the brain—or maybe because of it—it's not at all clear how the brain works. The same goes even for much simpler systems such as the eye or some molecular machines. It's hard to wrap one's mind completely around their workings, so it's comparatively easy to be talked out of one's strong initial impressions of design, especially by an authority figure in a lab coat. About the gears of Issus coeleoptratus, however, there's little ambiguity. The stark clarity of the structure is a standing rebuke to nonpurposive accounts of the system."*[58]

Concluding Thoughts

Is the ID argument devoid of issues or challenges? Certainly not. In fact, we will evaluate some of these challenges as we move forward in our text. However, the ID movement reveals that scientists can and often do rationally come to the scientific conclusion that there is a Designer, and, more specifically, that the Designer is God. Faith has developed as a result of evidence (cf. Hebrews 11:1), enabling believers to give respectable answers to honest inquiry.

As we close this chapter and prepare for the next, in which we will evaluate efforts to merge macroevolution with creationism, we conclude with two insightful quotes:

> *"There is an elephant in the roomful of scientists who are trying to explain the development of life. The elephant is labeled 'intelligent design.'*

---

[58]Ibid, 45-47.

To a person who does not feel obliged to restrict his search to unintelligent causes, the straightforward conclusion is that many biochemical systems were designed. They were designed not by the laws of nature, not by chance and necessity; rather, they were planned. The designer knew what the systems would look like when they were completed, then took steps to bring the systems about. Life on earth at its most fundamental level, in its most critical components, is the product of intelligent activity.

The conclusion of intelligent design flows naturally from the data itself—not from sacred books or sectarian beliefs. Inferring that biochemical systems were designed by an intelligent agent is a humdrum process that requires no new principles of logic or science. It comes simply from the hard work that biochemistry has done over the past forty years, combined with consideration of the way in which we reach conclusions of design every day."[59]

"Though the theory of intelligent design does not identify the agent responsible for the information—the signature—in the cell, it does affirm that the ultimate cause of life is personal. By personal I mean a self-conscious, deliberative mind in possession of thoughts, will, and intentions. Only persons have such minds and only minds of this kind can create complex specified information. If we know anything we certainly know this. Thus, while the theory of intelligent design does not prove the existence of God or answer all of our existential questions, it does reestablish the conditions of a meaningful 'search for meaning.' The case for intelligent design challenges the premise of the materialist credo and holds out the possibility of reversing the philosophy of despair that flows from it. Life is the product of mind; it was intended, purposed, 'previsioned.' Hence, there may be a reality behind matter that is worth investigating."[60]

---

[59]Behe, *Darwin's Black Box*, 193.

[60]Meyer, *Signature in the Cell*, 450-451.

Thought/Discussion Questions

1. Name three famous scientists who appreciated order in nature and praised our Creator.

2. Briefly describe Intelligent Design (ID) and its importance in the Creation/Evolution debate.

3. Is the "signature of design" a subjective or an objective argument? Is it persuasive?

4. What are some of the answers of evolutionary opponents to claims of divine design?

5. Provide two examples of intelligent design in nature.

6. Is ID sufficient of itself to inform us of the Creator? Explain your answer.

7. How does Hebrews 3:4 give a biblical support of ID?

What Are Evolution and Theistic Evolution?

In the preceding chapters, we have focused on highlighting the positive case for creationism as revealed in both general (i.e., nature) and special revelation (i.e., the Scriptures). We have also laid out the scriptural condemnation of atheism—a judgment that aligns with the observable natural order. To this point, we have only addressed the general theory of evolution in passing.

In this chapter, we will define the main components of evolutionary theory and then analyze the attempt to syncretize creationism and evolution (i.e., theistic evolution). In our subsequent assessment in Chapters 5-7, we will outline the multifaceted issues arising from amalgamating the two disparate worldviews. Then, in Chapters 8-10, we will explore the implications of the two systems as applied to the discipline of science and one's personal worldview. We aim to show the superior position of biblical creationism in light of the copious evidence available to humanity. But first, let's take a look at its chief competitor.

The General Theory of Evolution

The Father of Evolutionary Biology

In 1859, the English naturalist, geologist, and biologist Charles Darwin published *On the Origin of Species by Means of Natural Selection, Or the Preservation of Favoured Races in the Struggle for Life*. It was a watershed moment that resulted in radical scientific and sociocultural ramifications.[61] In the book, Darwin argued that biological diversity and disparity arose by common descent through a branching pattern of evolution through natural selection.

Darwin was not the first to posit methods of evolutionary biological development among biota; nevertheless, he was the most popular. Darwin's forbear-

[61]Many do not fully appreciate the widespread effects resulting from Darwin's evolutionary theory. For example, the influence of Social Darwinism in the late 19th and early 20th centuries led people like Adolf Hitler and Margaret Sanger to justify genocide and other atrocities on the premise of racial superiority.

ers included influential people like Jean-Baptiste Lamarck,[62] Charles Lyell,[63] and even his own grandfather Erasmus Darwin.[64] Each of these forebearers proposed varying systems dependent on spontaneous generation guided by uniform processes into greater complexity based on trait inheritance. Darwin's popularity was largely facilitated by the earlier publication of his journal kept from his five-year journey aboard the HMS Beagle. His celebrity and the influence of his scientific ideas burgeoned exponentially upon his publication of *The Descent of Man, and Selection in Relation to Sex* in 1871. After just a few decades, Darwin became the champion of evolutionary science.

Of course, Darwin's popularity was not the only factor in the explosive growth of his theory. Another key variable in this development was the militant atheistic outreach of Thomas H. Huxley and the X-Club in the 19th century to secularize science by promoting scientific naturalism. Consider, in this respect, the zeitgeist in the late 19th and early 20th centuries. The Enlightenment brought many to resist supernatural origin explanations for any and all phenomena. The sweeping religious revivals in the Protestant Restoration surely led to its converts, but these were philosophically doomed to be overshadowed by the post-God scientific revolution. For those weary of religious revivals, the theory provided an escape hatch not requiring appeal to the divine. Then, in the 1930s and 1940s, the modern evolutionary synthesis fully enthroned Darwinian Evolution.

---

[62]Jean-Baptiste Lamarck is most notable for his work in invertebrate zoology and taxonomy. He was a major proponent of evolutionary theory rooted in the inheritance of acquired characteristics. His most influential written works are *Système des animaux sans vertèbres* (1801) and *Philosophie zoologique* (1809).

[63]Charles Lyell is best known for his work in geology, most notably his popularization of uniformitarianism, which laid the groundwork for the concept of "deep time." His most influential work was *Principles of Geology: Being an Attempt to Explain the Former Changes of the Earth's Surface, by Reference to Causes Now in Operation,* published in three volumes in 1830-1833 by John Murray.

[64]Erasmus Darwin widely wrote in both prose and poetry regarding his own conception of evolution. His most influential contribution was his work *Zoonomia; or the Laws of Organic Life,* published in two volumes in 1794-1796. One of the most important quotes from the book is: "*Would it be too bold to imagine, that in the great length of time, since the earth began to exist, perhaps millions of ages before the commencement of the history of mankind, would it be too bold to imagine, that all warm-blooded animals have arisen from one living filament, which THE GREAT FIRST CAUSE endued with animality, with the power of acquiring new parts, attended with new propensities, directed by irritations, sensations, volitions, and associations; and thus possessing the faculty of continuing to improve by its own inherent activity, and of delivering down those improvements by generation to its posterity, world without end!*" (Section XXXIX.4.8).

Natural Selection

Darwin's chief argument lie in the foundational role of natural selection for evolution. So, what exactly was this core construct of evolutionary theory based upon? What evidence so convinced Darwin, and so many since then, that natural selection was the answer to other heavily debated evolutionary models of biological development? And what does "survival of the fittest" entail?

Ernst Mayr, one of the 20th century's leading evolutionary biologists, summarized natural selection using the following points:

- Every species is fertile enough that the population would grow if all offspring survived to reproduce.
- Despite periodic fluctuations, populations remain roughly the same size.
- Resources such as food are limited and are relatively stable over time.
- A struggle for survival ensues.
- Individuals in a population vary significantly from one another.
- Much of this variation is heritable.
- Individuals less suited to the environment are less likely to survive and reproduce; individuals more suited to the environment are more likely to survive and reproduce and leave their heritable traits to future generations, which produces the process of natural selection.
- This slowly effected process results in populations changing to adapt to their environments, and ultimately, these variations accumulate over time to form new species.[65]

Mutations, Reproduction, and Inheritance

Natural selection is not the only major component of the general theory of evolution. Another critical aspect of the theory is the incalculable amount of genetic mutations that have enabled forward development. Compounding this challenge, these mutations require survivability across reproduction as well as activation in progeny, itself a herculean difficulty that has consequently driven advancement in the field of epigenetics.

Ironically, natural selection undermines the very process of mutations driving evolutionary advancement as it serves, at least in theory, to *eliminate* the unnecessary baggage of biological systems. In many ways, natural selection seems best fit to devolve an organism by carefully trimming away the

---

[65]Ernst Mayr, *The Growth of Biological Thought* (Cambridge, MA: Harvard University Press, 1982), 479-480.

"extras" based on environmental and survival stipulations. This erosion effect has been the chief argument in the design movement in rebutting Darwinism advocates. Biochemist Michael Behe explains,

> *"As science rapidly advanced in the early twenty-first century, large studies showed only surprisingly minor changes in genes under severe selective pressure...now several decades into the twenty-first century, ever more sophisticated studies demonstrate that, ironically, random mutation and natural selection is in fact fiercely devolutionary. It turns out that mutation easily breaks or degrades genes, which counterintuitively, can sometimes help an organism to survive, so the damaged genes are hastily spread by natural selection. Strangely, in the space of a century and a half Darwinism has gone from the chief candidate for the explanation of life to a known threat to life's long-term integrity...although chance surely affects superficial aspects of biology, the newest evidence confirms that life is the intended work of a mind and that that work extends much more deeply into life than could previously be seen."*[66]

The Origin of Man (And Every Other Living Being)

An integral pillar of Darwinian evolution is that biological origins result from purely naturalistic processes. For this reason, the system obligates a starting point within the confines of the natural order. Enter the Big Bang theory, the bedfellow of Darwinism. Wedded together, this means that the general theory of evolution posits that everything we see and experience in the known universe began as a microscopic, energy-dense "seed" originating from the quantum fabric of space and exploded in a "big bang" that expanded over billions of years and organized into galaxies made of stars with planetary systems. Furthermore, life originated from matter and energy to form molecules, cells, more complex plants and animals, and eventually human beings in at least one planetary system. In layman's terms, the universe sprang forth by means of random chance. Nature evolves through eons of time, and biota, including humans, are plain products of the survival of the fittest. The materialistic combo seems to suggest that love, compassion, the common good, and the very concept of God do not objectively exist but are mere products of evolutionary advancement, which may or may not have occurred for the better. If cooperation exists at all, it is solely to propagate a particular gene pool: symbiosis for the sake of chance inheritance. This is far afield from the biblical portrait,

---

[66]Michael J. Behe, *Darwin Devolves: The New Science About DNA That Challenges Evolution* (New York: HarperOne, 2019), 10-11.

which speaks highly of the place of humanity as God's special creation apart from other biota (Genesis 1:26ff).

Icons of Evolution

Advocates for macroevolutionary theory appeal to various biological systems and alleged transitional forms to substantiate the theory. Darwin's journal entries detailing finch variation in the Galapagos Islands is a classic illustration. Dr. Bo Kirkwood comments,

> *"Of all examples in Darwin's evolutionary theory, it is the findings he observed on the Galapagos Islands, especially the finches, that are considered the strongest proof for it. These species of finches vary by size and coloring, and what Darwin primarily observed was the size and shape of their beaks, which seemed most suited for the type of environment and insects that occurred on the specific islands in which they were found. The finches then evolved into the specific species based upon natural selection and survival of the fittest. Interestingly, though Darwin certainly studied the Galapagos finches, they were never discussed in 'Origin of the Species.' They only became important many years later when those coming after Darwin elevated them to their present iconic status.*

1. Geospiza magnirostris 2. Geospiza fortis
3. Geospiza parvula 4. Certhidea olivacea
Finches from Galapagos Archipelago

> *Darwin's observations were unique for his time, but by now you understand that his finches in no way prove evolution, but only prove adaptation. Christians, as stated earlier, do not dispute microevolution or adaptation, but Darwin never observed a finch changing into another kind. In more recent studies, changes in the finches of the Galapagos Islands have been observed, and it is amazing how very quickly they can adapt to changes in the weather and other circumstances, but they clearly remain finches!"*[67]

Bernard Kettlewell's observations regarding color variation in peppered

[67]Bo Kirkwood, *Creation versus Evolution* (Athens, AL: Truth Publications, 2017), 57-58.

moths in England are another classic confirmation for theorists. Kettlewell observed two varieties of peppered moths and speculated that the darker variety flourished due to their darker color, which facilitated camouflage against the darker bark of birch trees. Many scientists have questioned his reasoning since then on the basis that peppered moths more typically avoid predators not by blending into birch bark but by resting underneath its leaves, more horizontal to the branches. Despite these challenges, some scientists and educational texts still offer the peppered moth as indisputable proof of evolutionary theory.[68]

Modern studies have directed attention to experiments involving quickly reproducing organisms such as bacteria in order to expedite mutation transference. While compelling and even groundbreaking at times, the quantum leap from these changes to the sorts of changes required to substantiate macroevolutionary theory is more than scientists could ever hope to bridge. Evil Knievel would have had an easier time jumping his motorcycle across the Grand Canyon!

Other icons of evolution depend on inferences drawn from similar body plans, genetic composition, and other comparisons. This is evident in alleged reptile-to-bird transitions such as archaeopteryx, sinosauropteryx, archaeoraptor, sinornithosaurus, etc. Evidential appeals are also made to alleged ape-to-human transitions such as Australopithecus afarensis, Neanderthal man, Cro-Magnon Man, Heidelberg Man, Nebraska Man, and other fossil discoveries. Both creationist and naturalist scientists discuss the plague of problems with each of these appeals.[69]

Inferences and Assumptions

Looking at the evidence, it should be clear that the general theory of evolution is not scientific law. It is an inference drawn from the available data based on the presuppositions of those often seeking an escape from the role of defendant before the judgment seat of Christ (cf. Romans 14:10; 2 Corinthians

---

[68]For example, Isaac Asimov's 884-page *New Guide to Science* has a mere half-page section on the evidence for Darwinism, citing the peppered moth example as sufficient to prove the whole theory.

[69]For more information on the plaguing problems of these alleged transitional forms, see: Eric Parker, *Behold the Builder: Scientific Evidences for the Biblical God* (Bowling Green, KY: One Stone Press, 2019), 61-75.

5:10).[70] Let's briefly look at some of the methodological presuppositions that so besmirch the noetic framework of those created in God's image.

Biological & Taxonomic Assumptions

The first major area of supposition relating to evolutionary theory involves biology and taxonomy. Consider several of the underlying assumptions of evolutionary theory, along with a few challenges to those assumptions:

- Non-living things gave rise to living material (i.e., abiogenesis or biopoiesis). This is in direct conflict with the scientific law of biogenesis, which requires life to produce life.[71]
- All living systems have a common biological origin (i.e., homology). However, we observe widely disparate systems with vastly different biochemical systems.
- Viruses, bacteria, protozoa, and higher animals are all interrelated. Yet, the evidence shows they are morphologically and physiologically different.
- Protozoa gave rise to Metazoa.
- The various invertebrate phyla are interrelated.
- The invertebrates gave rise to the vertebrates.
- Fish, amphibians, reptiles, birds, and mammals are interrelated. In stark contrast to this assertion, there is an obvious lack of transitional forms. This inference is also undone by observed stasis and discontinuity in the fossil record.[72]

Several secondary and tertiary assumptions spawn from this list. This system of assumptions caused many biologists to jettison the classical, fixed tax-

---

[70]Lest this accusation seem *ad hominem*, see the candid statements of several influential scientists in the work cited in the footnote above, specifically pages 163-165. Here is one example: "*We no longer feel ourselves to be guests in someone else's home and therefore obligated to make our behavior conform with a set of preexisting cosmic rules. It is our creation now. We make the rules. We establish the parameters or reality. We create the world, and because we do, we no longer feel beholden to outside forces. We no longer have to justify our behavior, for we are now the architects of the universe. We are responsible to nothing outside ourselves, for we are the kingdom, the power, and the glory forever.*" (Jeremy Rifkin, *Algeny* (New York: Viking Press, 1983), 244.)

[71]American chemist James Tour offers a free course rebutting abiogenesis. You may access that course on his YouTube page at: https://www.youtube.com/watch?v=WKLgQzWhO4Q

[72]These assumptions are a distillation of those listed in: G.A. Kerkut, "Implications of Evolution," in *International Series of Monographs on Pure and Applied Biology*, volume 4 (New York: Pergamon Press, 1960), 150-154.

onomies that took divine creation for granted (e.g., those of Carl Linnaeus and John Ray) in exchange for a great polyphyletic tree that traced humanity all the way down to primordial sludge. The original unrelated "kinds" of the Genesis creation account were exchanged for singular naturalistic origin. The implications for this evolution in taxonomy are pronounced and have influenced educational patterns and textbook illustrations across the board. In large part, the exchange was driven by the misunderstanding that the "kinds" of Genesis correspond to "species" when they truly correspond to higher taxonomic levels.[73] Moreover, manmade classification systems merely represent shared characteristics. They should not be understood to necessitate kinship. Unfortunately, these models create a chain of circular reasoning, which impacts fossil interpretation. This problem becomes even more skewed by faulted dating methods which depend upon many of the same assumptions (more on this in Chapter 5). These problems are all compounded by the ever-resilient retreat of evolutionists to the assumption that transitional forms will be uncovered in the fossil record at some point in the future, the same claim that Darwin offered to peer critics such as Louis Agassiz.

Philosophical Assumptions and Vocabulary Ambiguity

Another fundamental issue with Darwinian and neo-Darwinian theory is the jump from what is observable to what is unobservable. It is exceedingly common to observe extrapolation from the changes we can see and measure to those unseen and unmeasurable. This distinction in evolutionary theory must be considered in terms of distinguishing nomenclature such as "special" and "general" evolution or "microevolution" versus "macroevolution." G.A. Kerkut helpfully defined the difference in this way:

> *"There is a theory which states that many living animals can be observed over the course of time to undergo changes so that new species are formed. This can be called the 'Special Theory of Evolution' and can be demonstrated in certain cases by experiments. On the other hand, there is the theory that all the living forms in the world have arisen from a single source which itself came from an inorganic form. This theory can be called the 'General Theory of Evolution' and the evidence that supports it*

[73]What precisely the "kinds" of Genesis correspond to in current taxonomy systems is not clear. Genesis reveals that God made kinds, and reproduction beyond those kinds was not in His plan. Sometimes, "kinds" appears to be as specific as a single species (i.e., humans); other times, "kinds" is clearly in reference to a broader order.

> *is not sufficiently strong to allow us to consider it as anything more than a working hypothesis."*[74]

Kerkut later explains the systematic metaphysical and philosophical jumps from micro- and special evolution to macro- and general evolution:

> *"It might be suggested that if it is possible to show that the present-day forms are changing and the evolution is occurring at this level, why can't one extrapolate and say that this in effect has led to the changes we have seen right from the Viruses to the Mammals? Of course one can say that the small observable changes in modern species may be the sort of thing that lead to all the major changes, but what right have we to make such an extrapolation? We may feel that this is the answer to the problem, but is it a satisfactory answer? A blind acceptance of such a view may in fact be the closing of our eyes to as yet undiscovered factors which may remain undiscovered for many years if we believe that the answer has already been found.*
>
> *It seems at times as if many of our modern writers on evolution have had their views by some sort of revelation and they base their opinions on the evolution of life, from the simplest form to the complex, entirely on the nature of specific and intra-specific evolution. It is possible that this type of evolution can explain many of the present-day phenomena, but it is possible and indeed probable that many as yet unknown systems remain to be discovered and it is premature, not to say arrogant, on our part if we make any dogmatic assertion as to the mode of evolution of the major branches of the animal kingdom."*[75]

Phillip Johnson further discusses the assumptions that plague the very definition of evolution, as well as issues that can even arise in the binary distinction of micro- and macro-evolution. He states,

> *"Much confusion results from the fact that a single term—'evolution'—is used to designate processes that may have little or nothing in common. A shift in the relative numbers of dark and light moths in a population is called evolution, and so is creative process that produced the cell, the multicellular organism, the eye, and the human mind. The semantic*

---

[74]Kerkut, "Implications," 157.

[75]Ibid, 154-155.

implication is that evolution is fundamentally a single process, and Darwinists enthusiastically exploit that implication as a substitute for scientific evidence. Even the separation of evolution into its 'micro' and 'macro' varieties—which Darwinists generally resist—implies that all the creative processes involved in life comprise a single, two-part phenomenon that will be adequately understood when we discover a process that makes new species from existing ones. Possibly this is the case, but more probably it is not. The vocabulary of Darwinism inherently limits our comprehension of the difficulties by misleadingly covering them with the blanket term 'evolution.'"[76]

Theological Assumptions

One is hard-pressed to find a presentation of evolutionary theory without it saying something about theism. That is because the inferences used to jump from data to conclusion are riddled with anti-supernatural bias. If this bias could be dispelled and we could analyze the evidence with open minds, the issues of a *priori* atheistic bias could be put to bed. Michael Behe explains:

"When one starts to treat Darwinism as a hypothesis about the biochemical level of life rather than as an assumption, it takes about ten minutes to conclude it's radically inaccurate. It takes perhaps another ten minutes to realize that the molecular foundation of life was designed, and for effectively the same reason that Anaxagoras, Galen, and Paley reached the same conclusion for visible levels of biology (although, because of progress in science and philosophy, the argument is now necessarily much more detailed and nuanced than their versions): the signature of intelligent activity is the arrangement of disparate parts to fulfill some purpose. The molecular parts of the cell are elegantly arranged to fulfill many subsidiary purposes that must blend together in service of the large overall purpose of forming life…no unintelligent, undirected process—neither Darwin's mechanism nor any other—can account for that."[77]

You see, the evidence for evolutionary theory is often arrived at by abductive reasoning, which rules out the possibility of God before the evidence is even considered. D.M.S. Watson, the late zoologist of London University, ad-

---

[76]Phillip E. Johnson, *Darwin on Trial* (Washington, DC: Regnery Gateway, 1991), 69-70.

[77]Michael J. Behe, *Darwin Devolves: The New Science About DNA That Challenges Evolution* (New York: HarperOne, 2019), 8-9.

mitted, *"Evolution is a theory universally accepted, not because it can be proved true, but because the only alternative, Special Creation, is clearly impossible."*[78] Richard Lewontin, a Harvard geneticist, likewise admitted,

> *"We take the side of science in spite of the patent absurdity of some of the constructs. In spite of its failure to fulfill many of its extravagant promises of health and life, in spite of the tolerance of the scientific community for unsubstantiated just-so stories, because we have a prior commitment, a commitment to materialism. It is not that the methods and intuitions of science somehow compel us to accept a natural explanation of the phenomenal world, but on the contrary, we are forced by our prior adherence to natural causes to create an apparatus of investigation and a set of concepts that produce that materialism is absolute, for we cannot allow a Divine Foot in the door."*[79]

Can Evolution and Creationism Coexist?

Acknowledging the role of natural selection, the reality of mutations, and the truth of genetic inheritance, many believers have sought a partnership between evolution and creationism. This has been done mainly under the auspices of theistic evolution. Even for those who disparage evolution's explanatory power, there is a tendency to embrace the same timeline and advocate for deep time in which the forces of naturalism were leveraged by God in order to bring about His purposes (i.e., "progressive creationism"). The Roman Catholic Church is an excellent case in point as it has slipped toward this Gibeonite covenant.[80]

Despite this increasingly influential movement, this alignment runs contrary to the historical enmity of the two worldviews, as reflected in nearly every history of evolutionary theory. So, where does that leave today's believer? Is this partnership damnable heterodoxy or the honorable efforts of esteemed bridge-builders? Is the historical enmity merely a repeat of the infamous dissension of Galileo and the Roman Catholic Church? Is there room for joining

---

[78]D.M.S. Watson, "Adaptation," *Nature* 124:233, 1929.

[79]Richard Lewontin, "Billions and Billions of Demons," a review of *The Demon-Haunted World: Science as a Candle in the Dark* by Carl Sagan, 1997, *The New York Review* (January 9, 1997), 31.

[80]Pope Pius XII's encyclical of 1950, *Humani generis*, was a major step in this approval. Other developmental steps include Pope John Paul II's address to the Pontifical Academy of Sciences on October 22, 1996, and Pope Francis' address to the Pontifical Academy of Sciences on October 27, 2014. The Gibeonites were a group of Canaanites who tricked Israel into a covenant by leveraging a pronounced lack of circumspection on Israel's part (Joshua 9).

the two systems, or do they retain irreconcilable differences?

Believers should begin by evaluating whether or not the Bible's information content with respect to creation accounts is sufficiently broad enough to permit this affinity. Devotion to the doctrine of biblical sufficiency in equipping believers unto all good works, as well as the doctrine of divine inspiration, preclude bridge-building atop the carcasses of sound doctrine (1 Corinthians 2:7-16; 2 Timothy 3:16-17; Jude 3). The Bible repeatedly extols the sanctity of accurately preserving God's inspired Word, a collection of documents containing numerous references to God's purposeful creation of the universe. Paul thus encouraged Timothy, *"Be diligent to present yourself approved to God, a worker who does not need to be ashamed, rightly dividing the word of truth"* (2 Timothy 2:15). This was due to the fact that in latter days, men would compromise sound doctrine and heap up teachers in accordance with their own desires (cf. 2 Timothy 2:23-26; 3:1-5; 4:1-5). The one who does not accurately handle God's teachings but chooses to add to or take away from the inspired Word is identified as an infidel, accursed of God (Deuteronomy 4:2; 2 John 9-11; Revelation 22:18-19). As we will examine in the next few chapters, one must repeatedly violate the sanctity of the written Word in order to accept macroevolutionary theory. As such, a firm distinction must be preserved; otherwise, one compromises the integrity of the faith and embraces error (cf. 2 Corinthians 6:14-7:1).

The second important consideration for believers is whether or not the observable phenomena of general creation can simultaneously correspond with both the creationist and evolutionist positions. If so, then a harmonization may be reasonable and, indeed, responsible. As we have already mentioned, this is made difficult by the breadth of denotations and connotations attached to the term evolution and its various constituents. As mentioned earlier, there are definite truths to many of the primary components of evolutionary theory. Biological changes do take place in populations; those changes can be shared across generations, and environment plays a role in the development of those changes. Those are undeniable, observable facts. In addition to that, the world and universe look very old, and the assumption of uniformitarianism has some merit based on comparative observations. However, there are colossal issues that do not permit the jump from divinely designed adaptations to full acceptance of all evolutionary tenets, especially including origins.

Whether or not we can combine evolution and creationism is of vital importance because either the universe was created by an omnipotent, omniscient, and omnipresent Creator, or it wasn't. Either He created the world largely as it stands today, with disparate kinds of animals and plants reproducing after their kinds and humanity assigned with His image from the start, or the Bible relies on Mesopotamian-like mythological accounts to describe

Twelve Ways That Theistic Evolution Compromises the Creation Account
1. Adam and Eve were not the first human beings (and perhaps they never even existed).
2. Adam and Eve were born from human parents.
3. God did not act directly or specially to create Adam out of dust from the ground.
4. God did not dirctly create Eve from a rib taken from Adam's side.
5. Adam and Eve were never sinless human beings.
6. Adam and Eve did not commit the first human sins, for human beings were doing morally evil things long before Adam and Eve.
7. Human death did not begin as a result of Adam's sin, for human beings existed long before Adam and Eve and they were always subject to death.
8. Not all human beings have descended from Adam and Eve, for there were thousands of other human beings on Earth at the time that God chose two of them as Adam and Eve.
9. God did not directly act in the natural world to create different "kinds" of fish, birds and land animals.
10. God did not "rest" from his work of creation or stop any special creative activity after plants, animals, and human beings appeared on the earth.
11. God never created an originally "very good" natural world in the sense of a world that was a safe environment, free of thorns and thistles and similar harmful things.
12. After Adam and Eve sinned, God did not place any curse on the world that changed the workings of the natural world and made it more hostle to mankind.
This list is taken from: Wayne Grudem, "Theistic Evolution Undeermines Twelve Creation Events and Several Crucial Christian Doctrines," in *Theistic Evolution: A Scientific, Philisophical, and Theological Critique* edited by J.P. Moreland, Stephen C. Meyer, Christopher Shaw, Ann K. Gauger, and Wayne Grudem (Wheaton, IL: Crossway, 2017).

God's use of a naturalistic methodology to bring about humanity over generations and generations of failed experiments. Either we believe in one God as Creator of the universe, intervening in it and sustaining a personal relation to His creatures, or we believe in chance processes in which God is prey to unintelligent, mechanistic forces of nature. Believers must accept God's Word as a permanent, perfect message to us as His children with all of its history, teachings, and consequences. We cannot pick and choose (cf. Matthew 6:24).

Concluding Thoughts

Believers can be deceived into thinking that their faith is unassailable (1 Corinthians 10:12). The Apostles often made that same mistake, and Jesus regularly admonished them for their meager faith (e.g., Matthew 8:23-26; 14:22-33). Faith is frequently diluted by fear, and the purity of sound doctrine is often polluted by the trapping philosophies of humankind (Colossians 2:8). Satan knows this and has perfected the art of leading humanity into religious and philosophical syncretism (Galatians 5:7-9).

Having examined the chief competitor of creationism, we want to direct our attention in the next three chapters to three core issues of theistic evolution. Namely, we will look at differences in origin, timeline, operation, and purpose. Keep the backbone of these first four chapters in your mind as we reveal the ramshackle scaffolding that will collapse under the feet of those believers who stand for the false doctrine of theistic evolution.

Thought/Discussion Questions

1. How important was Darwin to the acceptance of evolutionary theory?

2. What role do mutations and natural selection play in evolutionary thought? Are these two functions consistent with the Bible?

3. What is the difference between the general theory of evolution and theistic evolution? Is there a difference?

4. Identify a few icons of evolution.

5. What are some of the inferences and assumptions that evolutionists make?

6. What does the Bible say about conforming the Scriptures to fallible human ideas?

7. Can evolution and creationism coexist? Is theistic evolution tenable?

CHAPTER 5

Creation vs. Evolution

The most divisive subject of the creation/evolution debate relates to origin and timeline. This conflict is even pronounced among believers. This is primarily because what the Bible describes and what we observe does not always seem to agree with one another. That has led many to compromise the integrity of the Scriptures in an effort to harmonize the two sources of information. In this chapter, we will focus on the core issues at stake in this debate, summarizing irreconcilable origin and timeline issues between creationism and evolution, addressing the abortive attempts of theistic evolutionists to synchronize the views, offering resolution to the supposed conflicts between what the Bible states and what humans observe, and concluding with why acknowledging the differences between the viewpoints matters.

Comparing Evolutionary and Creationist Origins and Timelines[81]

First, let's consider origins conflicts. The contents of Genesis 1 guide the biblical model for origins. The opening line states, *"In the beginning God created the heavens and the earth."* This unforgettable opening ties the origin of everything in the universe to God, identifying Him as the only self-existing and autonomous force (cf. Exodus 3:14-15). He is the Fountain, Author, and Source of all life (Psalm 36:9; John 1:4; 5:26; Acts 3:15). As the psalmist reminds God's supplicants: *"Know that the Lord Himself is God; It is He who has made us, and not we ourselves; We are His people and the sheep of His pasture"* (Psalm 100:3). The Bible then establishes and describes a timeline encompassing thousands of years, tracing the events of history back to original creation in which God created all things (excepting Himself) in six days with a seventh sanctified day devoted to restful reflection. Genesis accomplishes this through the use of narrative depictions of creation in conjunction with both linear and non-linear genealogies, which bridge between original creation and the historical patriarch Joseph.[82] From this timeline, we can then trace forward to the

[81]For a quick video on major discrepancies between the evolutionary and creationist timetables, see: Eric Lyons, "Creation Versus Evolutionary Chronology," Apologetics Press, November 9, 2020, https://apologeticspress.org/video/creation-versus-evolutionary-chronology-or-time-evolution-and-the-bible-5892/

[82]The Hebrew *toledot* formula represents this genealogical backbone for Genesis.

modern era, thus providing a firm chronology restricted to thousands of years.

Contrast the alternate model of origins presented in the general theory of evolution. In the evolutionary model, the universe is spontaneously generated by mere serendipity. Chance and probability rule throughout known history, including the evolutionary development of people. The primary basis for evolution's multi-billion-year timeline is the necessity of epochs of time to enable the incalculable number of adaptations needed to develop modern species.

Despite the obvious conflict in the systems, theistic evolutionists (and other advocates of Old-Earth creation) hope to reconcile the issue through creative interpretations of Genesis 1. Before we get more into that, let's establish a critical four-point outline for each model's dissonant origin views for the universe, the Earth, life, and humankind.

The Origin of the Universe

As aforementioned, the biblical model for the creation of the universe is heavily contingent on the opening statement of the Bible. The creation of *"the heavens and the earth"* envelopes all creation into a uniform, harmonized, and good product of the Creator. The eternality of God and His desire to beget a creation that would magnify His glorification combined with His omnipotence to generate our universe out of nothing. The timeline overtly places the creation of the universe in near proximity to the creation of the Earth, all planetary systems, all kinds of biota, and the progenitor and progenitrix of humanity.

The general theory of evolution, often adjoined to the big bang theory, describes the origin of the universe as self-generated out of quantum foam, originating from a supposed energy field that is the fabric of space-time. This expanded from unimaginable density to eventually result in the evolution of galaxies, stars, molecules, and life across vast expanses of time employing either gradual or punctuated developments. The current generally accepted age of the universe, as we know it, is 13-14 billion years. The non-biblical astrophysical evidence presented for this timeline primarily relies on the observed expansion of the universe and the cosmic microwave background. It merits stating that this timeline wildly fluctuates between different models and has also exponentially increased over the last two centuries.

Creationists and evolutionists look at the universe and interpret its origin and development in two distinctive ways. General evolutionists say that the universe appears a certain age and that those processes took a vast amount of time to occur. Creationists interpret the universe as created in a particular, functional configuration by the Creator. Any specific age of the universe is

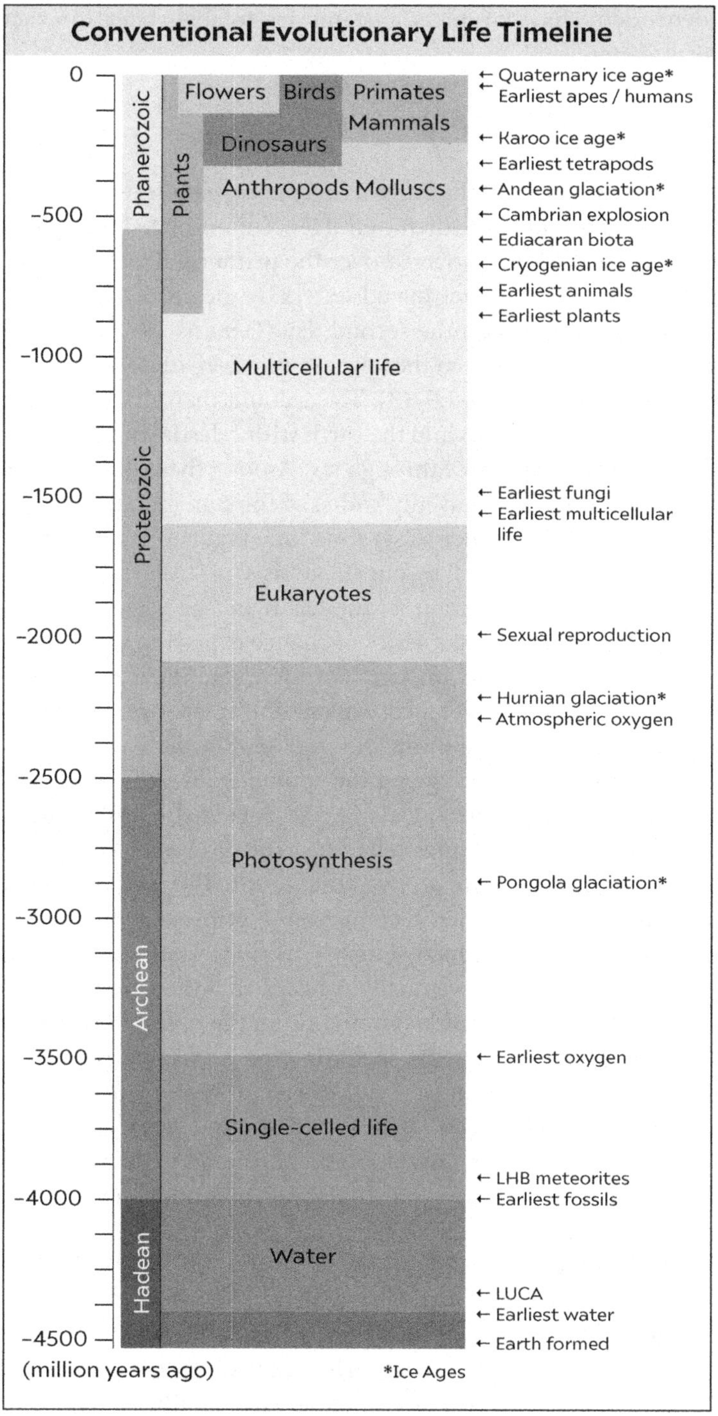
Conventional Evolutionary Life Timeline
0
-500
-1000
-1500
-2000
-2500
-3000
-3500
-4000
-4500
(million years ago)
Phanerozoic
Proterozoic
Archean
Hadean
Plants
Flowers
Birds
Primates
Mammals
Dinosaurs
Anthropods Molluscs
Multicellular life
Eukaryotes
Photosynthesis
Single-celled life
Water
← Quaternary ice age*
← Earliest apes / humans
← Karoo ice age*
← Earliest tetrapods
← Andean glaciation*
← Cambrian explosion
← Ediacaran biota
← Cryogenian ice age*
← Earliest animals
← Earliest plants
← Earliest fungi
← Earliest multicellular life
← Sexual reproduction
← Hurnian glaciation*
← Atmospheric oxygen
← Pongola glaciation*
← Earliest oxygen
← LHB meteorites
← Earliest fossils
← LUCA
← Earliest water
← Earth formed
*Ice Ages

interpretation-dependent since no material stamp irrefutably states an exact birthdate for the universe.

The Origin of the Earth and Solar System

The Bible describes the origin of the Earth as occurring parallel with the origin of the universe. In particular, the divine effulgence illuminated the *"formless and void"* creation as the Spirit hovered over the primordial waters (Genesis 1:1-5). God then divided the illuminated waters by means of an expanse into the upper and lower regions on the second day (Genesis 1:6-8). On day three, God caused dry land to appear in the lower waters, a vivid description of the generation of the Earth (Genesis 1:9-13). The account then shifts to God's immediate population of the heavens and the earth with celestial bodies, flora, fauna, and humanity across days three through six. Among these descriptions is the intentional formation of the heavenly bodies of the sun, moon, and stars after the creation of the Earth in order to serve as time regulators and to provide natural light upon the Earth on day four (Genesis 1:14-19).

The evolutionary model for the origin of the Earth and its related celestial bodies is markedly different. In short, a series of chance explosions, collisions, compressions, and cooling resulted in their cosmic evolutionary materialization. For those seeking a more detailed explanation, there are a series of hypothesized models that culminate in the development of the Earth and its sister bodies in the solar system. Following the spontaneous generation and expansion of the universe from the Big Bang, the solar nebula hypothesis claims that a large, rotating cloud of interstellar dust and gas began to contract due to the shockwave of a nearby supernova. This resulted in the collapse of the nebular center, followed by immense compression until its hydrogen began to fuse into helium. This forged a T Tauri star, which later evolved into our sun. The evolution of our sun eventually resulted in a theorized process known as runaway accretion. This resulted in the clumping of large fragments of dust and debris into planetary systems, including the Earth, approximately 4.5-5 billion years ago. This process of runaway accretion resulted in our planetary systems and is postulated to have taken 10-100 million years. Sometime following that proposed development of the Earth, our moon perhaps formed from the Earth's collision with a large proto-planet sometimes called Theia (i.e., the Giant-impact hypothesis).

Side-by-side, this is what we see. Cosmic evolution alleges that the sun and other heavenly bodies evolved millions of years before the Earth. In contrast, Genesis 1 has a water-covered Earth created on day one (1:1) and the celestial bodies made on day four (1:16). It is inescapable that we cannot have both. Cosmic evolution alleges that billions of years following the Big Bang, Earth evolved out of a massive cloud of dust billions of miles wide. There was also no water on the surface of the early Earth, as bodies of water did not form for millions of years. In contrast, God spoke a water-covered Earth into existence on the first day of creation (1:1). On day two, He divided the waters (1:6). Again, we cannot have both.

The Origin of Life

The biblical paradigm develops the special creation of all life forms immediately following the establishment of the Earth. Plants are the first living things, created on the third day (Genesis 1:11-12). Their position as the lowest biological forms is clear from their expressly stated purpose to serve as food for higher life forms, although they also possess aesthetic value (Genesis 1:29-30; 2:8-9, 15-17). Note also that God created all plant kinds together, designed with reproductive ability according to their kinds. The second life forms created are twofold – winged animals and sea creatures – all with reproductive means within their respective kinds, created on day five (Genesis 1:20-23). Day six recounts God's work in creating cattle, creeping things, and beasts of the earth, all according to their kinds. The pinnacle of day six is the creation of the highest life form, humanity. Although God created all mankind on the same day as the last grouping of animals, there is a noticeable shift in the unique depiction of mankind being created in God's image, thus distinguishing him from beasts.

In the evolutionary model for the origin of life, the primordial Earth became habitable through the eventual step-by-step cooling of its temperature, development of continents, dawn of the oceans, and formulation of a hospitable atmosphere, all transpiring during the Hadean and early-Archean Eons. Around 3.7 billion years ago, early life forms arose from the gradual, chance organization of primordial pandemonium. In this model, simpler organic compounds like nucleotides and amino acids led to single-celled life in the early-to-mid Archaean Eon. Then, in the subsequent Proterozoic Eon, the earliest forms of multicellular life and fungi arose. In the next period, the Phanerozoic Eon, plants, arthropods, mollusks, dinosaurs, mammals, flowers, birds, primates, and eventually humans arose. Evolutionary scientists have offered countless models of abiogenesis and even extraterrestrial seeding of our planet (e.g., directed panspermia), which led to simpler life forms evolving step by step into modern species across billions of years, culminating in all contemporary flora and fauna. A stark contrast with the biblical paradigm is that evolution contends that birds arose from land animals. In contrast, the Bible has birds created first and land animals created a day later. Other differences include evolution's view of some fish and insects arising before the development of fruit-bearing trees.[83]

The Origin of Humanity[84]

The biblical origins model implicitly "saves the best for last," assigning humanity's creation as the apex of God's original creative work. Following the creation of land beasts early on the sixth day, the text highlights mankind's creation:

> *Then God said, "Let Us make man in Our image, according to Our likeness; let them have dominion over the fish of the sea, over the birds of the air, and over the cattle, over all the earth and over every creeping thing that creeps on the earth." So God created man in His own image; in the image of God He created him; male and female He created them. Then God blessed them, and God said to them, "Be fruitful and multiply; fill the earth and subdue it; have dominion over the fish of the sea, over the birds of the air, and over every living thing that moves on the earth."*
>
> *And God said, "See, I have given you every herb that yields seed which is on the face of all the earth, and every tree whose fruit yields seed; to you it shall be for food. Also, to every beast of the earth, to every bird of the air, and to everything that creeps on the earth, in which there is life, I have given every green herb for food"; and it was so. Then God saw everything that He had made, and indeed it was very good. So the evening and the morning were the sixth day.*

There is an unmistakable emphasis in the creation account of Genesis 1—humanity's creation in the image of God. No other kind is given that divine signature. Additionally, mankind is given the creation mandate, being placed as authority over all living things. This truth also has a bearing on the ability of humans to kill and eat animals post-Flood (cf. Genesis 9:1-7). Genesis 2 builds on the creation of man in a few more ways. God specifically made Adam from the dust of the earth, and breathed into his nostrils the breath of life. In so doing, Adam became a living being (Genesis 2:7). God placed Adam into the Garden of Eden to keep and cultivate the garden (Genesis 2:8-17). God then

---

[83]For a succinct summary of the gradual evolutionary development of plants, see: James O'Donoghue, "A Forest is Born," *New Scientist* 196[2631]:38, November 24, 2007.

[84]This section is indebted to: Eric Lyons, "Man and the Age of the Earth," Apologetics Press, November 9, 2020, https://apologeticspress.org/video/man-and-the-age-of-the-earth-5886/

deemed that Adam would excel with a suitable helper and created Eve from one of Adam's ribs (Genesis 2:18-25). Biblically, humanity is not the great evolutionary advancement of the animals; humanity is separate, apart from, and superior to all animals.

Contrast the evolutionary model of human development. As animals gradually became more developed, primates arose several million years ago. From the primates, evolutionary advances slowly formed semi-human creatures referred to as hominids. Those hominids became more and more advanced based on specific mutations and natural selection, leading to early humans like *homo erectus*. Eventually, *homo sapiens* arose approximately 300,000 years ago. In theistic evolutionary models, God supernaturally consigned souls and His image to these evolutionary ancestors at some point in this latter period. This is, of course, *ad hoc* and indefensible, but that is the overall state of the evolutionary history of humans. In short, evolutionary theory contends that the origin of humanity is separated from the cosmic order by about 13.9997 billion years. This flies in the face of the biblical portrait, which places God's goals for humanity (John 17:24; 1 Peter 1:20; Ephesians 1:4) and their actual establishment at the start of the cosmic order (Mark 10:6; Luke 11:50-51; Romans 1:20). Clearly, there are irreconcilable differences between the views that theistic evolution cannot explain away!

Bungled Attempts to Synchronize the Worldviews

Side-by-side, creationism and evolution have vastly divergent timelines and views on origin. Despite these observable discrepancies, some have sought harmonization of the views to advocate for theistic evolution. This is primarily due to the apparent age exhibited by the Earth and the universe at large. As such, the desire to harmonize the account in Genesis with what we observe is understandable, though terribly faulted and detrimentally consequential. These proposed reconciliations do a great disservice to the biblical text in an effort to conform the Bible to deficient human conclusions arrived at by faulted human methods of observation and testing. We will briefly examine a few primary methods employed to blend the views and highlight some major problems with those methods.

Before we do that, though, we want to state something very clearly: God could have created the world however He desired. If God decreed that His desire was to use evolution to accomplish the development of kinds over vast periods, He would be unimpeachably free to do that. Additionally, suppose Genesis and other passages of Scripture left unspecified a timetable by which to arrive at conclusions regarding God's creation of the world. In that case, we might be free to conclude that God used evolution and long spans of time to accomplish His creation. Such possibilities would, perhaps, not terribly infringe upon the scheme of redemption available to humanity in Jesus Christ (although the differing value and salvation of an ape-to-man transition would be a hard truth to swallow). This would be true not only of the theistic evolutionist's view but also of other non-evolutionary old-Earth creation models, such as progressive creationism. The problem, though, for these theorists, is that the Bible and many facets of creation point to a young Earth and universe. Likewise, a straightforward reading of Genesis 1-2 and other creation Scriptures do not require adding vast amounts of time—interpreters have to insert that into the text. Nevertheless, let's consider a few models that attempt to produce reconciliation.

Gap Theory and the Multi-Gap Theory

One widely accepted pathway to insert eons of time into the Genesis creation account is to interject an indeterminate gap of time between Genesis 1:1 and 1:2. Some have even offered arguments from Hebrew grammar that certain words and phrases should be understood and/or translated in such a way as to more clearly convey this alleged time gap. One of these arguments is a proposed shift from "*was formless and void*" to "*became formless and void*." The idea being that God first made an amorphous Earth before slowly developing it either directly or indirectly. Another argument requires understanding a distinction between the different Hebrew words for creation employed in the text (*bara* and *asah*). Regarding the latter contention, while there are nuances between the terms, they are often used interchangeably for God's creative activity (e.g., Genesis 2:4; 5:1; 6:7; Isaiah 41:20; 43:7; 45:12, 18; Amos 4:13). Moreover, there are at least thirteen different Hebrew terms the Scriptures use for God's creation, so this facile distinction in terms is misleading. Some have even advocated for more than one gap in the text with non-consecutive days separated by long periods of time between each "day" of creation. Famously influential proponents of gap theory include George H. Pember and John Clayton.[85]

Despite the fanciful resorts to "the original Hebrew," which preys on the ignorance of laymen, the most natural and straightforward reading of Genesis 1 neither requires nor points to any unmentioned gaps in the text. The proposed linguistic arguments have all been sufficiently rebutted by Hebrew scholars. Moreover, this theory relies on what the Bible does not say rather than what

---

[85]For Pember, see: George H. Pember, *Earth's Earliest Ages* (London: Hodder and Stoughton, 1907), reprint. For Clayton, see: John N. Clayton, *The Source: Eternal Design or Infinite Accident?* (Niles, MI: Clayton, 2011); also, John N. Clayton, *The Rational God: Does God Make Sense?* (Niles, MI: Clayton, 2015).

The Hebrew Vocabulary of Creation		
Hebrew Term	**Select Scriptural Passage (s)**	**English Translation**
ברא/*bārā'*	Deutronomy 4:32; Malachi 2:10	"created"
עשה/*'āsāh*	2 Kings 19:15; Jeremiah 32:17	"made"
יצר/*yātsar*	Isaiah 43:7; 45:18	"formed"
פעל/*pā'al*	Exodus 15:17; Proverbs 16:4	"made"
שים/*sîm*	Exodus 4:11; Proverbs 8:29	"fixed, appointed"
כון/*kūn*	Isaiah 45:18; Jeremiah 10:12	"established"
יסד/*yāsad*	Psalm 24:2; Proverbs 3:19	"founded"
בנה/*bānāh*	Genesis 2:22; Amos 9:6	"built"
ילד/*yālad*	Deuteronomy 32:18; Job 38:29	"begat"
חול/*chūl*	Job 15:7; Psalm 90:2	"brought forth"
נטה/*nātāh*	Job 9:8; Isaiah 42:5	"stretched out" (heavens)
קנה/*qānāh*	Deuteronomy 32:6; Psalm 139:13	"formed" (literally "bought")
עבד/*'āvad*	Jeremiah 10:11; Daniel 6:28 (Aramaic)	"made"

it does say, a terribly suspect method of interpretation. One Hebrew scholar summarizes the state of gap theory this way:

> *"There is nothing in the Hebrew text of Genesis 1 to demand a gap of time. The Hebrews in fact had a variety of ways to express chronological gaps, whether general or specific. For general amounts of time they could and often did say, 'after this' (acharēy-kēn) or 'after these things' (acharēy-haddevarîm hā-'ēleh). To express a greater extent of time, they could have said 'many days' (yāmîm rābbîm) or something similar. Although common in the Bible, none of these phrases occurs in Genesis 1. So we are left to trust the Gap theorists that they are qualified to speak where the Bible is silent, and to understand in the Hebrew what no Hebrew*

scholars actually affirm, and what no qualified translators have ever put forth. So is Gap Theory linguistically viable? No." [86]

Day-Age Theory [87]

Often found in tandem with gap theory is the invocation of day-age theory, which argues that each "day" of Genesis is an indeterminate period that can account for the prolonged developments of the evolutionary model. The argument is grounded in the semantic range of the Hebrew word for day (singular: *yom*; plural: *yammim*); namely, that "day" can refer not only to the 12-hour period of light or to the 24-hour period of time but also to an unspecified amount of time (e.g., "the day of the LORD"). While there is enough semantic range to include an indeterminate amount of time by means of *yom/yammim,* as the Old Testament uses the terms over 2,300x, this is a highly misleading reduction of the data. First, in the narrative genre, in which Genesis is primarily presented, a day is regularly a reference to a 24-hour period. The alternative usage of a broad period or age is more evident in prophetic and poetic genres like Psalms (e.g., 90:4) or Joel (e.g., 2:28-32). Second, the days of creation are interpreted as 24-hour periods by Mosaic standards such as Sabbath observance (Exodus 20:11; 31:17; et al.). Third, biblical context is straightforward to establish what denotation is at play in the use of day, akin to our own modern usage. The days of Genesis are specifically enveloped with contextual clues as to their 24-hour reference, most notably the time reference of *"an evening and a morning."* One lexicographer explains, *"The words 'morning' (Hebrew boqer) and 'evening' (Hebrew 'ereb) appear a total of 348 times in the Old Testament, and they are used in reference to specific, defined portions of a regular 24-hour day."*[88] Fourth, the use of numerals and ordinals in reference to yom/yammim all point to normal 24-hour usage (cf. Genesis 8:10-11; Exodus 18:13; Numbers 9:21). Henry Morris clarifies, *"Whenever a limiting numeral or ordinal is attached to 'day' in the Old Testament (and there are over 200 such instances), the meaning is always that of a literal day."*[89] Fifth, if "days" are not a limit-

---

[86]Justin Rogers, "Is the Gap Theory Linguistically Viable?" *Reason & Revelation* 35.12 (December 2015), https://apologeticspress.org/wp-content/uploads/2021/08/1512.pdf

[87]For a brief, helpful overview of this theory and its problems, see: Eric Lyons, "The Day-Age Theory," Apologetics Press, October 1, 2021, https://apologeticspress.org/video/the-day-age-theory-5893/

[88]A.H. Konkel, *New International Dictionary of Old Testament Theology & Exegesis,* ed. Willem A VanGemeren (Grand Rapids: Zondervan, 1997), 1:711, 716.

[89]Henry Morris, *Scientific Creationism* (San Diego: Creation-Life Publishers, 1974), 224.

ing time reference to 24-hour periods, then what does the reference "years" in Genesis 1:14 point to? To this question could be added other challenging problems, such as: How was plant life able to survive millions of years in total darkness? How did plants survive millions of years without the key role of pollinators, which God did not create until "days" 5-6?

The day-age theory is an artificial effort to conform the Bible with a faulted understanding of the apparent age evident in the world, a built-in feature intentionally and purposefully designed to accomplish certain divine goals for the world. These divine goals included population expansion, biodiversification through sexual reproduction, and reliable time regulation. The addition of vast amounts of age offers no true reconciliation of the differences between the evolutionary and Genesis models. There are also other discrepancies, including a very different order of development. This arbitrary addition of time does not redeem major order differences in astrophysical and biological development.

Mythologizing Biblical Creation Accounts

The biblical text simply does not support the efforts of gap and day-age theorists. This has led many to abandon these theories and to advocate for a total re-reading of the Genesis creation account as a myth comparable to ancient Near Eastern (ANE) creation myths. While there are interesting parallels between these creation accounts, ANE myths are best understood as echoes of the original creation passed down and enhanced by oral tradition. This is especially the case as the Bible creation account reads much differently as to genre and narrative in Genesis 1-2, universally condemns understanding biblical truth as myth (cf. 1 Timothy 1:3-7; 4:6-10; 2 Timothy 4:1-5; Titus 1:10-16; 2 Peter 1:16-21), and regularly refers to the Genesis account as factual history (e.g., Deuteronomy 32:8; 1 Chronicles 1:1; Job 31:33; Matthew 19:4; Luke 3:38; Romans 5:12-21; 1 Timothy 2:13ff).[90]

Resolving Apparent Conflicts [91]

If the answer is not to be found in interpolating eons of time into the Genesis account, how do we account for the apparent age evident in the created order? What do we make of chronological methods, geological structures, and

[90]For more information on the Bible's view of myths, see: Eric Parker, *Behold the Record: Historical Evidences for Christianity* (Bowling Green, KY: One Stone Press, 2023), 49-73.

[91]For further information on this subject, see: Dr. Don DeYoung, *Thousands…Not Billions* (Green Forest, AR: Master Books, 2008).

biodiversification, which seem to all point to colossal amounts of time? These scientific measures point to the biggest difficulty in making sense of the Genesis account of creation. Christians should be aware of the main talking points that enable profitable discussions. These can, in turn, lead to a more robust theology and the conversion of souls who conscientiously struggle with this conflict.

What Exactly Does "Old Age" Look Like?[92]

To begin with, the question of age depends on a few assumptions, primarily the need for comparison. An older adult is understood in contrast to a young person. Similarly, an ancient document is contrasted with a new document. As applied to understanding the age of the Earth, what precisely is our means of comparison? What does it mean to be billions of years old? What exactly does that look like? Does anyone actually know? Who was a witness to the Earth's birth certificate? Can we compare the age of the Earth with other planets? How can anyone reliably say, "The Earth looks billions of years old?" Compared to what exactly? This surface-level conclusion depends a whole lot on comparison and epistemological data that is patently unavailable. Ultimately, a subjective view of apparent age relies on a paradigmatic worldview through which information is filtered and interpreted. Since evolutionary theory demands an old Earth, it should be no surprise that many people cannot help but see the Earth as billions of years old. That claim is not so much based on the data as on philosophical worldview and confirmation bias.

Dating Methods and Their Representation

But what about more objective and scientific assessments of data? Not just the subjective look of apparent age, but objective measurements that indicate an old Earth? Truthfully, there are scientific measurements that point to an old Earth, so Christians should equip themselves with specific knowledge to respond to these points (Proverbs 15:28; 1 Peter 3:15). These matters constitute the essence of the creationist/evolutionary dichotomy in understanding the age of the Earth and its corollary matters.

One primary element of the debate relates to ages determined by radiometric dating. Radiometric dating takes modern samples of things like lead and examines the past emission of alpha- and beta-particles from radioactive parent elements like uranium. As the original parent element (e.g., uranium)

[92]This section is indebted to: Eric Lyons, "Miracles and Earth's Apparent Age," Apologetics Press, November 9, 2020, https://apologeticspress.org/video/miracles-and-earths-apparent-age-5895/

in the soil broke down, it eventually became a daughter element (e.g., lead). Radiometric dating traces this pattern of decomposition based on half-life signatures and extrapolates a relative age signature from the data. While there are great benefits to these methods of dating, there are also complications that can sorely impact interpretation. Radiometric dating methods rely on three key assumptions:

- How much of the element there was at the beginning (i.e., the zero-date problem);
- A uniform decay rate over time (i.e., uniformitarianism); and
- Little-to-no cross contamination of the sample (i.e., the leakage problem).

This means that radiometric dating techniques can often lead to circular reasoning, especially due to the wholesale acceptance of evolutionary theory. There are also often-unmentioned differences between measuring the dating of rocks and the dating of fossils since completely fossilized samples cannot be dated by any radioactive methods. As a result, geological chronologists will date nearby igneous rocks to date a fossil based on the assumption that they were co-occurring, but igneous rocks can and often do form in ways that defy radiometric dating. Another significant issue with this dating method is that there can be a wide disparity across the measurement of different elements, leading to dates varying up to many millions of years.

Geology is a definite area where chronologies come into conflict. In addition to radiometric dating, there are also geological time questions about canyon erosion, the development of fossil beds, the maturation of fossil fuels and diamonds, petrification, stratification, ice core dating, the development of lake varves, dendrochronology (i.e., dating based on tree ring development), the development of caves and speleogenesis (i.e., the formation of stalagmites and stalactites), mountain formation, and tectonic plate movement. Old-Earth geologists often use measurements from these areas to develop the geologic column. Rarely do they note that the popularly accepted geologic column is a compendium of data from various sources into one theoretical model. It is nowhere observed in totality; if it were, the strata layer would need to be over one hundred miles deep! To visualize that, consider that the deepest canyon that we can navigate without drilling is the Grand Canyon. The Grand Canyon is only a mile and a half deep at its deepest point. In short, the geologic column is an interpretation of data and not itself a fact.[93] Besides the absence of the entire

[93]While faulted, the geologic column is not without some benefits. Oil companies, for example, regularly use it to look for oil because most oil deposits are found in the "Jurassic" layer.

theoretical geologic column, there are also interpretational issues of circular reasoning dependent on evolutionary theory. This is especially evident in the presupposition of uniformitarianism, the belief that all geological processes have operated in the past in the same way they now function in the present. This presupposition impacts the interpretation of index fossils, the timeline of dinosaurs, and a host of other matters. Consequently, if uniformitarianism is proven suspect, all of these purportedly rock-solid evidences come into question, at least as to their interpretation.

Before we address uniformitarianism and its philosophical competitor, catastrophism, we want to address fossils in more detail. When Darwin presented his model for evolution and natural selection acting on mutations became vogue, his theory lacked the support of the fossil record, an issue brought to his attention by peers such as Louis Agassiz. Darwin resorted to his own "god of the gaps" and claimed that time would reveal transitional forms among fossils waiting to be uncovered. Time has passed, and innumerable fossils have been found, but not the ones Darwin needed to support his theory! The fossils we have uncovered do not represent the Darwinian tree of biological development, do not explain the early origin of phyla, do not represent simple life forms from the start, and do not offer transitional forms. Instead, there is full formation and observed complexity at the earliest fossil layers (e.g., the Cambrian layer), disparity predating diversity, and fossil stasis in which biota retain their form and either continue in that form or go extinct. We also find evidence of rapid water erosion and sedimentary deposition in which nests, embryos, and the movement and activities of animals at the time of those often swiftly occurring events are evident. Fossils and other discoveries have also illustrated the co-occurrence of dinosaurs and humans, offering evidence for a vastly different timeline than evolution, which requires a 65-million-year gap between the two! These fascinating fossil finds and other chronological markers point to a young Earth timetable in corroboration with the literal reading of the Genesis creation account. [94]

Uniformitarianism and Catastrophism

We have noted at several junctures that chronological methods are impacted by uniformitarianism, the idea that current geological processes precisely reflect what has always been the case. The trailblazer of this scientific presupposition was Charles Lyell. There is some degree of truth to this assumption

[94]For more detailed information on the relationship between fossils, dinosaurs, and young Earth creationism, see: Eric Parker, *Behold the Builder: Scientific Evidences for the Biblical God* (Bowling Green, KY: One Stone Press, 2019), 45-90.

since we have begun measuring geological processes in the past few hundred years. Geological processes do tend to follow measurable patterns. Nevertheless, there is a competing idea that argues that catastrophic processes in the past can impact our interpretation of geological evidence. This view is known as catastrophism and integral to the creationist's case for a young Earth. To be clear, catastrophism does not challenge the *tendency* for uniform geology patterns; rather, catastrophism argues that geological processes do not *always* follow the same patterns. The question is whether or not there is reason to accept the postulated impacts of catastrophism and how much bearing that has on our interpretation of the available evidence.

There are several good reasons to accept catastrophism. For instance, catastrophic events in modern times often lead to results that an unwitting scientist would explain in terms of long-age development. Volcanoes, earthquakes, floods, and other cataclysmic events can radically change landscapes and profoundly impact dating methods.[95] Canyon development, river erosion, speleogenesis, cave development, ice layering, fossilization, carbonization, petrification, and a host of other key markers for age have all been shown to be expedited by catastrophic processes in the present.[96] Caves can form more quickly via sulfuric acid instead of slower carbonic acid. Speleothems, usually produced one inch per century, can develop at different rates depending on rain saturation, differing surface air temperature, drip rate, droplet concentration, and carbon dioxide levels in the soil. Severe weather also impacts measures such as dendrochronology and ice-core dating. All of these matters have been replicated in laboratories that mimic catastrophic processes and reveal the errors of uniformitarian assumptions.[97]

The key historical testimony to a catastrophic impact on fossils is the biblical account of the worldwide flood of Noah as recorded in Genesis 6-9, especially in such dramatic landscape changes as noted in 7:11-12 where *"all the fountains of the great deep were broken up."* Most of the oil, coal beds, and fossil graveyards in the Earth, which many contend are evidence of an old Earth,

---

[95]The rapid development of the Paricutin Volcano starting in 1943 is an excellent example of how catastrophic forces can incredibly change a landscape. Another classic example is the 1980 eruption of Mount St. Helens.

[96]Notable examples of rapid canyon development include Burlingame Canyon in 1926, the Glen Canyon Dam cavitation in 1983, the Little Grand Canyon in 1982, Canyon Lake Gorge in 2002, the Coralville Lake Gorge in 1993, and Providence Canyon since the early 1800s. For more information, see: Jeff Miller, "Do Canyons Prove an Old Earth?" Apologetics Press, November 19, 2020, https://apologeticspress.org/video/do-canyons-prove-an-old-earth/

[97]Jeff Miller, "Have Evolutionists Proved That the Earth Is Billions of Years Old?" Apologetics Press, November 11, 2020, https://apologeticspress.org/video/have-evolutionists-proved-that-the-earth-is-billions-of-years-old-5936/

can be and have been, rationally explained as a result of the worldwide deluge of Noah's day. So, even if the world did "look" old, people cannot discount or deny the impact of catastrophes which can explain this appearance in conjunction with original divine purposes.

God's Intentional Design Requires the Appearance of Age [98]

The impact of catastrophes has much bearing on our modern interpretation of the available evidence in order to ascertain the timeline of our origins. We must also be open to considering other additional variables. For instance, that the Earth may in some ways appear older than it is can be explained in light of the nature of God's miracles. The Bible often portrays God bypassing the uniform order to accomplish His goals (e.g., John 2:1-11; Matthew 14:13-21). In the same way, Genesis tells us that God created fruit-bearing plants, not just seeds, and fully-grown birds, not just eggs. He immediately made full-grown trees that would have had many tree rings if cut down. He created a full-grown man capable of walking, talking, working, and procreating. If one can philosophically accept it, God miraculously made a mature creation by means of His supernatural power.

With this in mind, let's consider a specific challenge to young Earth creationism: the distant starlight problem. The challenge asserts that since some stars are an estimated fifteen billion light years away, evolutionists assume that the universe must be at least close to fifteen billion years old. This is furnished as astrological proof for an old Earth and is part of the main motivation for theistic evolutionists to merge the timelines. However, consider God's intentional purpose and design for the stars to illustrate His handiwork and immediately provide a regular scale by which humans could understand time (Genesis 1:14-19). Believers should have no qualms accepting God's creation of both stars and their light rays in a way that they instantaneously function for their intended purpose.[99]

---

[98]This section is also indebted to: Eric Lyons, "Miracles and Earth's Apparent Age," Apologetics Press, November 9, 2020, https://apologeticspress.org/video/miracles-and-earths-apparent-age-5895/

[99]Other explanations have been offered for the resolution of the distant starlight problem, including gravitational time dilation (GTD) and alternate synchrony conventions (ASC). For more information, see: Jason Lisle, "Anisotropic Synchrony Convention—A Solution to the Distant Starlight Problem," *Answers Research Journal* vol. 3 (2010): 191-207), https://answersresearchjournal.org/anisotropic-synchrony-distant-starlight/#

Do Timing, Chronology, and Order Really Matter? [100]

"What's the big deal? Does it really matter? We can still know that God exists, the Bible is His Word, God created the universe at some point in the past, Jesus came and died for our sins, one day He will return and judge the world, etc. The focus of the gospel is salvation in Jesus, not the age of the Earth." This is a common response to this detailed problem between evolution and creationism. While we do not always know the reasons behind someone saying this, there are serious problems with this response.

First, the Bible implicitly and explicitly teaches that God's creation is young. God tailor-made creation on His revealed timetable, and He is clear about what that timetable looks like. The Holy Spirit provides time references in the creation account as well as time references elsewhere (e.g., genealogies in Genesis 5, 10-11; 1 Chronicles 1-9; Matthew 1; Luke 3; etc.) by which a dating of thousands of years is fixed.[101] Had he not done that, the door may be open to an old Earth; however, He has revealed a timetable and order of events, so we should respect that. One commentator passionately states,

> *"So far as I know there is no professor of Hebrew or Old Testament at any world-class university who does not believe that the writer(s) of Genesis 1–11 intended to convey to their readers the ideas that (a) creation took place in a series of six days which were the same as the days of 24 hours we now experience; (b) the figures contained in the Genesis genealogies provide by simple addition a chronology from the beginning of the world up to the later stages of the Biblical story, and (c) Noah's flood was understood to be worldwide, and to have extinguished all human and land animal life except for those in the ark"* [102]

Sometimes, on the part of Christians, there is an unwillingness to examine the arguments, which are often intense and technical. This is unhelpful and only plays into the caricature of Christians being anti-academic, backward,

---

[100]Eric Lyons, "Why Address the Age of the Earth?" Apologetics Press, November 9, 2020, https://apologeticspress.org/video/why-address-the-age-of-the-earth-5887/

[101]Some genealogies in the Bible are linear, transitioning directly through one generation. Other genealogies are non-linear for a variety of reasons but still within measurable limits. For example, Genesis 5 is a linear genealogy. Genesis 11 is more non-linear, but the lower and upper limits of the record would still be 390-4,800 years, given up to a 900-year lifespan.

[102]Alvin Plantinga, "Evolution, Neutrality, and Antecedent Probability: A Reply to McMullin and Van Till," in *Intelligent Design Creationism and Its Critics: Philosophical, Theological and Scientific Perspectives*, ed. Robert T. Pennock (Cambridge: MIT Press, 2001), 217.

and unyielding to cognitive development. When this is the case, Christians miss good opportunities to personally grow and also to help others overcome obstacles in coming to faith (Philippians 2:15). Another immoral motivator is fear and shame (Romans 1:16; 2 Timothy 1:7; 1 John 4:18). Many believers feel overwhelmed, or ashamed, to take a stand for biblical truth that runs contrary to common opinion or accepted thought. Feelings of inferiority, or the fear that faith is ill-grounded and not in agreement with scientific knowledge, cause some to retreat to the intellectual fetal position. That's no good either and ends in the same place as our first issue. No Christian should be ashamed of anything that Almighty God has revealed in His Word, including the age and development of the Earth.

Third, a firm stand for the creationist timeline and order of creation refutes evolutionary theory. We mentioned above that evolution has thoroughly impacted science and seemingly every other field of knowledge as well. Evolutionary theory has detrimentally impacted economics, sociology, and a host of other learning disciplines. In fact, evolutionary theory has been a major catalyst in lower views of objective truth, leading to the rise of moral relativism and postmodernism. Defense against this dark and diabolical corruption of God's purpose for humanity is a must for conscientious Christians, and we have the tools available to dismantle evolutionary theory. Michael Le Page admits this when he states, *"A young Earth would...be a problem for evolution, since evolution by natural selection requires vast stretches of time—'deep time'—as Darwin realized."* [103]

Fourth, compromise on the first pages of the Bible inevitably leads to compromises in later sections of the Bible. Atheistic attacks on the dating of the exodus, the reign of David, the historical resurrection of Jesus, and other pivotal events in the Bible become grounds to question the Bible in its entirety without reasonable consideration of adequate responses to such attacks, all available within the same fields in which these attacks are advanced. Popular creationist Ken Ham explains:

> *"As soon as you surrender the Bible's authority in one area, you 'unlock the door' to do the same thing in other areas. Once the door of compromise is open, even if ajar just a little, subsequent generations push the*

---

[103]Michael Le Page, "Evolution: A Guide for the Not-Yet Perplexed," *New Scientist* 198[2652]:24-33, April 19, 2008, 26.

> *door open wider. Ultimately, this compromise has been a major contributing factor in the loss of biblical authority in our Western world."*[104]

Concluding Thoughts

In the days of King Ahab of Israel, the prophet Elijah challenged the people of Israel to stop faltering between two opinions (1 Kings 18:21). Their compromise had brought a division between their nation and their God. Likewise, Jesus admonished the early churches at Pergamum, Thyatira, and Sardis for their spirit of compromise (Revelation 2-3). Concession seems to be a timeless struggle for God's people. We are in the same predicament of faltering between irreconcilable creationist and evolutionary timetables. To be true to our Creator and His Word, there can be no faithful harmonization of these divergent models.

Hebrew scholar Justin Rogers speaks to the widespread influence and danger of this compromise when he warns that,

> *"Many Christians working in the field of scientific cosmology seek to poach godless theories from modern science and work them into a model of biblical faith. We should applaud their efforts so long as they do not 'go beyond what is written' (1 Corinthians 4:6). Unfortunately, some do go too far. In their desire to harmonize the biblical account with the scientific 'necessity' of old-Earth creationism, they seek to read into the Bible concepts not clearly present. Rather than using divine inspiration to inform science, they prefer to impose modern scientific insight onto the Bible—an insight, it should be observed, the original readers of the Bible would not have understood."*[105]

Christians need to take a stand for the biblical timeline and its view of origins. Disciples of Jesus Christ need to have a backbone (cf. Joshua 24:15). While theistic evolutionists naively pursue and advocate for the faulted path of compromise, the facts prevent this unholy union. One can either accept the evolutionary timeline and its view of origins or the Bible and its view of origins. It's option A or option B; there is no option C.

---

[104]Ken Ham, *The New Answers Book Volume 4: Over 30 Questions on Creation/Evolution and the Bible* (Green Forest, AR: New Leaf Publishing Group, 2013), 18.

[105]Justin Rogers, "Is the Gap Theory Linguistically Viable?" *Reason & Revelation* 35.12 (December 2015), https://apologeticspress.org/wp-content/uploads/2021/08/1512.pdf

Thought/Discussion Questions

1. Why is the overall timeline of such importance in the Creation/General Evolution debate?

2. Compare and contrast the biblical and the evolutionary timelines.

3. Compare and contrast the biblical and evolutionary origins of the universe.

4. Explain the origin of life in the creationist and evolutionist frameworks.

5. What are some of the differences between the biblical and evolutionary views of humanity?

6. Explain some of the bungled attempts to synchronize evolutionary and biblical timelines.

7. Identify some of the apparent difficulties the creationist faces between the Bible's timeline and what we see in nature today.

CHAPTER 6

Creation vs. Evolution – Variation

In the last chapter, we saw that the two time paradigms of macroevolution and biblical creationism cannot coalesce. The long-range timeline necessary for macroevolution, itself a problem for biblical creationists, also requires the methodology of combinatorial variations arising through genetic mutations preserved by natural selection, a subject introduced in Chapter 4. The big question is: can the mindless teamwork of time, genetic mutations, and natural selection *actually* produce the sweeping variation we observe today? And an ancillary question for theistic evolutionists, did God empower this trio of nature to produce the astounding variety present in our world apart from His direct creative hand in ways imperceptible to humanity? To answer these questions, this chapter will examine the reality of variations as well as the limits of unguided natural variation over time. Along the way, we will analyze the explanatory power of evolution versus creationism, discussing how the numbers become prohibitive to exclusively naturalistic explanations for adaptation.

The Reality of Variation

One of the great wonders of life is the inestimable amount of variation we see across our planet. This is not an oblique or hidden phenomenon. All people of all belief persuasions recognize this truth. Enormous variation is unmistakable within subjects of a single species and across all other classification levels. The immensity of distinction, diversity, and disparity evident in our world is palpable. This marvel is further magnified by the presence of fossils that reveal the average rate of extinction of around 200 species of flora and fauna per year, exponentially compounding the number of adaptations and changes.

Major figures in the biblical narrative noted these differences and strove to classify plants and animals in ways that systematized human knowledge of such differentiations (e.g., Adam, Job, Solomon, etc.). Modern scientists build on these efforts and, despite our many advances, struggle to categorize all living things (i.e., taxonomy) definitively. One primary challenge for taxonomists involves the degree of deviation necessary to justify a separate classification. Just how much variance is enough to warrant a new label? A second difficulty is that certain plants and animals defy labeling. Whales and bats are both mammals but are obviously very different. Other animals with complicated histories of classification include the fossa, sunda flying lemur, gerbil, platypus, maned wolf, star-nosed mole, and pangolin.[106] Some organisms, like sea anemones, look like plants but are actually animals. Some plants, like the Venus flytrap, don't seem to act like plants would, devouring animals for food. Some animals, like the sea slug, can borrow chloroplasts and store the plant pigment chlorophyll for use as camouflage. The blue dragon slug takes it further, enslaving whole algal cells! A third complication for taxonomists involves balancing categories to prevent oversaturation or undersaturation of single classifications. At what point does a category get "too full," justifying division into smaller categories? At what point does a category become too restrictive? The occurrence of innumerable variations leads to countless questions about classification and the "why" behind such differences.

The Evolutionary Explanation for Variation

How does an evolutionist begin to explain the "why" behind these differences? The genius of Darwinian evolution is its elegant simplicity in explaining biodiversity, at least in proposal. Small, slight changes accrue over time, leading upward to the plethora of complex life forms we see today. Tree of life models, primate-to-man representations, feather evolution mockups, and other illustrative tools for evolutionary biology are readily palatable. Who could disagree with a system as straightforward and appealing as that? Humans regularly observe small changes snowballing into major changes outside of biology; should we not expect this in biology, too? Maybe, with enough time, minute changes can transform non-living chemicals into the marvelous biological complexity we see today?

This sounds unpretentious, even incontrovertible, but on closer examination, the proposed methodology requires innumerable factors, an exquisitely serendipitous chain of events, and fanciful extrapolation from limited data (Proverbs 18:17). The questions begin to multiply quickly as thinkers peek behind the curtain.

- Where did the material for these changes originate?
- How did these slight variations arise?
- How were the changes preserved across time?
- How did mindless forces "know" to preserve advantageous factors and dispense of disadvantageous factors?

---

[106]Examples derived from: "Unclassifiable Animals that Challenged Biologists," OpenMind BBVA, 15 March 2022, https://www.bbvaopenmind.com/en/science/bioscience/unclassifiable-animals-that-challenged-biologists/

- How did simpler living things survive long enough without specific later systems like blood clotting already in place?
- How did sexual reproduction enter the picture?
- What guarantees are there that the systems which now exist are the best?
- What guarantees do humans have that our cognitive faculties are reliable?
- Is the human mind really just a matter of naturalistically developed neurological circuits?
- Why are certain behaviors that counteract evolutionary advancement, like homosexuality and transgenderism, still so pronounced today?
- Why is there such diversity at all and not humankind exclusively, assuming of course that humanity even is the highest life form?

These questions, and many others besides, reveal that the theory of evolution is infinitely more complicated than the one-dimensional models often offered. So, what are the proposed answers of evolutionary scientists?

A Starting Point?

Let's start with the origin of information. Darwin famously admitted the complexity of the human eye befuddled evolutionary explanation but then proceeded to beg a starting point. In his request, he asked his readers to "suppose" that we "assumed" the existence of a simpler sight mechanism. Then, having been granted this starting point, he extrapolates to an increasingly more complex eye. But that isn't really a good place to start, is it? Evolutionary scientists regularly demand such an allowance. This is reminiscent of the old joke about human arrogance that goes,

> *One day, a group of scientists got together and decided that humanity had come a long way and no longer needed God. So they picked one scientist to go and tell Him they were done with Him. The scientist approached God and said, "God, we've decided that we no longer need you. We're to the point where we can clone people, manipulate atoms, build molecules, fly through space, and do many other miraculous things. So why don't you just go away and mind your own business from now on?"*
>
> *God listened very patiently and kindly to the man. After the scientist was done talking, God said, "Very well. How about this? Before I go, let's say we have a human-making contest." To which the scientist replied, "Okay, we can handle that!"*

"But," God added, "we're going to do this just like I did back in the old days with Adam."

The scientist nodded, "Sure, no problem," and bent down and picked up a handful of dirt. God wagged a finger at him and said, "Uh, uh, uh. Put that down. Please go find your own dirt."

This prerequisite also calls to mind the now popular pithy phrase, *"Natural selection may explain the survival of the fittest, but it cannot explain the arrival of the fittest."*[107] Evolution really does struggle with origins, an obvious irony in light of the titles emblazoned upon the published works of its greatest pioneer! This is true for biological evolution and its proposed precursor, chemical evolution. Evolutionary theory requires DNA, RNA, and protein to develop the first cell, but protein is necessary to make DNA! How do we add information without information to start with? Werner Gitt summarizes the need for information from the start and how that requires a programmer in this seven-part way:

- There can be no information without a code.
- Any code is the result of a free and deliberate convention.
- There can be no information without a sender.
- Any given chain of information points to a mental source.
- There can be no information without volition (will).
- There can be no information unless all five hierarchical levels are involved: statistics, syntax, semantics, pragmatics, and apobetics.
- Information cannot originate in statistical processes.[108]

---

[107]Andreas Wagner is best known for the modern use of this phrase. It serves as the inspiration for some of his most influential published works, including: Andreas Wagner, *Arrival of the Fittest: How Nature Innovates* (New York: Penguin Random House, 2014).

[108]Werner Gitt, *In the Beginning Was Information: A Scientist Explains the Incredible Design in Nature* (Green Forest, AR: Masters Books, 2007), 80.

Amino acid	Genetic code	Abbr
Alanine	GCA GCC GCG GCU	Ala
Arginine	AGA AGG CGA CGC CGG CGU	Arg
Asparagine	AAC AAU	Asn
Aspartic acid	GAC GAU	Asp
Cysteine	UGC UGU	Cys
Glutamine	CAA CAG	Gln
Glutamic acid	GAA GAG	Glu
Glycine	GGA GGC GGG GGU	Gly
Histidine	CAC CAU	His
Isoleucine	AUA AUC AUU	Ile
Leucine	CUA CUC CUG CUU UUA UUG	Leu
Lysine	AAA AAG	Lys
Methionine	AUG	Met
Phenylalanine	UUC UUU	Phe
Proline	CCA CCC CCG CCU	Pro
Serine	AGC AGU UCA UCC UCG UCU	Ser
Threnine	ACA ACC ACG ACU	Thr
Tryptophan	UGG	Try
Tyrosine	UAC UAU	Tyr
Valine	GUA GUC GUG GUU	Val
Stop sign	UAA UAG UGA	

Natural Selection and Environmental Constraints

But let's grant a starting point to Darwin and his students. Where do evolutionary scientists proceed from there? Observing slight variation, such as differing beak size among finches and colored variation in peppered moths, suggests that small changes develop over time in response to environment. Fair enough; in fact, quite agreeable. These are certainly not the only adaptations scientists or non-scientists have observed. Such observed changes are slight, though. The finches are still finches. The moths are still moths. At this point, evolutionary scientists point to similarities in body plans and functions across species. The pentadactyl configuration for instance, or structural patterns underlying scales, the epidermis, and feathers. These observed similarities across categories lead to the philosophical leaps of homology. Small changes could lead to more extensive changes, thus explaining these similarities. But what would be the driving force behind these changes?

Enter natural selection. It is readily observable that food needs, population changes, and other environmental factors lead to adaptations. Specific traits can be amplified in a species in a particular ecosystem. Moreover, there can be perceived changes in the physical characteristics of a species in a given ecosystem. However, even though physical characteristics (e.g., color, markings, height, speed, etc.) may appear to change over time in a given population in a specific locale, the kind of organism does not change. This level of adaptation has never been observed by a human on planet Earth, in a laboratory setting or otherwise. While environmental constraints may draw out certain characteristics already preprogrammed into biota, this does not constitute evidence for macroevolution. In fact, it does not even mean the organism is the best of its classification. As Gary Parker summarized,

> *"Being the fittest, then, is no guarantee of survival at all. It may only mean you are likely to be the last of your kind to die out! Fitness has to do with competition within a group; survival of the group often depends on competition among different groups, often related to changing environmental factors – loss of habitat, increase or decrease in temperature or moisture levels, changes in the saltiness of aquatic and soil environments, catastrophes like fires, floods, earthquakes, underwater landslides, etc."*[109]

The Critical Role of Positive Mutations

Perhaps, though, natural selection could leverage genetic anomalies and mutations to facilitate these macroevolutionary adaptations? The hypothesis of positive genetic mutation leading to genetic drift quickly became integral to evolutionary science. Genetics seemed like a panacea for all the problems of how environmental constraints could lead not only to minor adaptations but lasting and magnificent changes. Through advances in genetics, the propositions of Darwin, for many, have become indomitable. The instrumental work of scientists like Gregor Mendel paved the way for eventually mapping the human genome, understanding genetic sequencing, and studying genetic mutations. Mendel also showed what can happen through crossbreeding different species, both positively and negatively. The artificial breeding of roses, dogs, horses, and pea plants all pointed to the possibility that nature could perhaps lead to natural selection. The offspring of such experiments have interesting and useful traits, such as a beautiful rose or an incredibly fast racehorse; however, roses still only produce roses, and horses still only produce horses (or sterile offspring such as a mule, the hybridization of a horse and donkey).

Moreover, artificial selection requires a mind; natural selection requires no mind to guide the process, even within the theories of theistic evolutionists.

---

[109]Gary Parker, *Creation Facts of Life: How Real Science Reveals the Hand of God* (Green Forest, AR: Master Books, 2006), 89.

Biologists often emphasize this limited application of genetic variation. One source states,

> *"Starting in the 1970s, many biologists began questioning its adequacy in explaining evolution. Genetics might be adequate for explaining micro-evolution, but microevolutionary changes in gene frequency were not seen as able to turn a reptile into a mammal or to convert a fish into an amphibian. Microevolution looks at adaptations that concern the survival of the fittest, not the arrival of the fittest. As Goodwin (1994) points out, 'the origin of species—Darwin's problem—remains unsolved.'"*[110]

Think for a moment, too, about what mutations actually are. When geneticists find a mutation in a baby's DNA, how do the baby's parents react? What does this mean? For example, geneticists discover that a child has Trisomy 21, a chromosomal mutation resulting from an extra copy of the 21st chromosome. Will this make the child's life and the parent's lives easier or more difficult? No doubt, a diagnosis of Down Syndrome would be less than preferable.[111] There is a reason mutations have a negative connotation attached. Overwhelmingly, even when extra information is added to the genome, the result introduces difficulties and issues. Yes, mutations can elicit genetic variation, but this is overwhelmingly catastrophic to survival and sexual reproduction.

Additionally, natural selection regularly prunes these anomalies from the genome and disposes of these aberrations as the species reproduces. This brings us back to the cosmological issues of evolution; namely, mutations are not genetic scriptwriters. They are typographic errors in a genetic script that is already written. As Gary Parker observes, *"Most mutations are caused by radiation or replication errors...the gene has to be there before it can mutate. All you get as a result of mutation is just a varied form of an already-existing gene, i.e., variation within kind."*[112]

---

[110]S.F. Gilbert, J.M. Opitz, and R.A. Raff, "Resynthesizing Evolutionary and Developmental Biology," *Developmental Biology* 173 (1996), 361. The quote is taken from: Brian Goodwin, *How the Leopard Changed Its Spots: The Evolution of Complexity* (New York: Scribner, 1994).

[111]Perhaps this goes without saying with the audience of this book, but this is not to say that a challenging diagnosis such as Down Syndrome is hopeless or that it only brings challenges. By the grace of God, such a diagnosis can deepen the faith of the parents and open many doors to effective service for the Lord. Moreover, the Scriptures teach the dignity of all humans, despite any attendant challenges resulting from mutations in a sin-laden world.

[112]Parker, *Creation Facts of Life*, 118.

Mutations and Prohibitive Numbers[113]

Despite the connotation of negativity attached to mutations, evolutionism has led many to enthrone mutations. Again, Gary Parker helpfully points this out by saying,

> *"Uncritical acceptance of evolution has so stunted scientific thinking that people give mutations god-like qualities. They act as if a cosmic ray striking a cell can cause a 'mutation' that somehow assembles over 1,500 DNA bases into a brand-new gene, regulators and all, and suddenly begins producing a brand-new protein responsible for a brand-new trait, raising the lucky mutated organism to the next higher rung on the evolutionary ladder! ... Typically, a mutation changes only one letter in a genetic sentence averaging 1,500 letters long."*[114]

The simple models portraying evolution vastly undersell the computational and statistical complexities behind such propositions. As Parker points out, mutations are not god-like, are very minor changes, and rarely prompt positive benefits. Others, like Douglas Axe, have done research to put numbers on these difficulties. Axe conservatively estimates a 1 in 10^{77} rarity of the generation of novel genetic information sufficient to produce novel protein folds utilizing the mutation-natural selection mechanism.[115] Other scientists have presented similar statistics.[116] One Christian evidences book records these corresponding thoughts:

> *"The probability of the chance development of a very simple system*

---

[113]While broader than this discussion, there are helpful resources which explain the mathematics supporting creationism and Christianity in general. One such example is: Marvin L. Bittinger, *The Faith Equation: Mathematical Evidence for Christianity*, 2nd ed. (Santa Fe, NM: Advantage Inspirational, 2022). While this resource has some compromises in respect to timeline and the role of evolution, it does offer beneficial information.

[114]Parker, *Creation Facts of Life*, 120.

[115]Douglas D. Axe, "Estimating the Prevalence of Protein Sequences Adopting Functional Enzyme Folds," *Journal of Molecular Biology* 341.5, August 27, 2004, 1295-1315, https://doi.org/10.1016/j.jmb.2004.06.058

[116]Reidhaar-Olson, John, and Robert Sauer. "Functionally Acceptable Solutions in Two Alpha-Helical Regions of Lambda Repressor." *Proteins: Structure, Function, and Genetics* 7 (1990): 306-16; Yockey, Hubert P. "A Calculation of the Probability of Spontaneous Biogenesis by Information Theory," *Journal of Theoretical Biology* 67 (1977c): 377-98; Yockey, Hubert. "On the Information Content of Cytochrome C," *Journal of Theoretical Biology* 67 (1977b) 345-376.

composed of only 200 integrated parts is 1 out of 10^{375} (this is the number 1 with 375 zeroes after it!). Mathematicians generally consider any event with a probability of less than 1 chance in 10^{50} as having a zero possibility; in other words, they say it's impossible.

A 200-part system is considered a mathematical possibility, yet it falls far short of any living system we know. The most basic type of protein molecule that could be classified as living is composed of at least 400 linked amino acids, each made up of a specific arrangement of four or five chemical elements, each with a unique combination of protons, neutrons, and electrons.

The chance formation of even the simplest replicating protein molecule is 1 in 10^{450}. The probability of forming the proteins and DNA for the smallest self-replicating entity is 1 in $10^{167,626}$."[117]

Genes and Base Pairs: A Comparison		
Species	**Genes (estimated)**	**Base Pairs**
Yeast *(Sacchoromyces cerevisice)*	6,000	12 million
Round worm *(Coenorhabitis elegons)*	19,000	99 million
Mouse *(Mus musculus)*	40,000	3 billion
Fruit Fly *(Drosophila melanogaster)*	13,600	165 million
Human *(Homo sapiens)*	40,000	3 billion

Many evolutionists, like Richard Dawkins, wave their hands at these prohibitive statistics.[118] Such scoffers claim that probability is irrelevant; however, it provides perspective to the overly-simplistic charts and schemes presented in defense of evolution. Some evolutionists are more honest with these statistics and appeal to more specific mutations in epigenetic structures, Hox genes,

[117]Heath Rogers, *Introduction to Christian Evidences* (Bowling Green, KY: One Stone Press, 2017), 25.

[118]One example of Dawkins' dismissal may be found at: https://whyevolutionistrue.com/2016/03/20/god-vs-physics-krauss-debates-meyer-and-lamoureaux/#comment-1316386

or developmental gene regulatory networks (dGRN's), but all of these appeals have met with even worse statistical likelihoods.[119]

The Bible, "Kinds", and Divine Oversight

The problems for the evolutionary explanation of biological variation are manifold. The mathematics behind such possibilities helps to expose this reality. Microevolution is observable, and organisms can respond to their en-

A Conventional Polyphyletic Tree of Darwinian Evolution

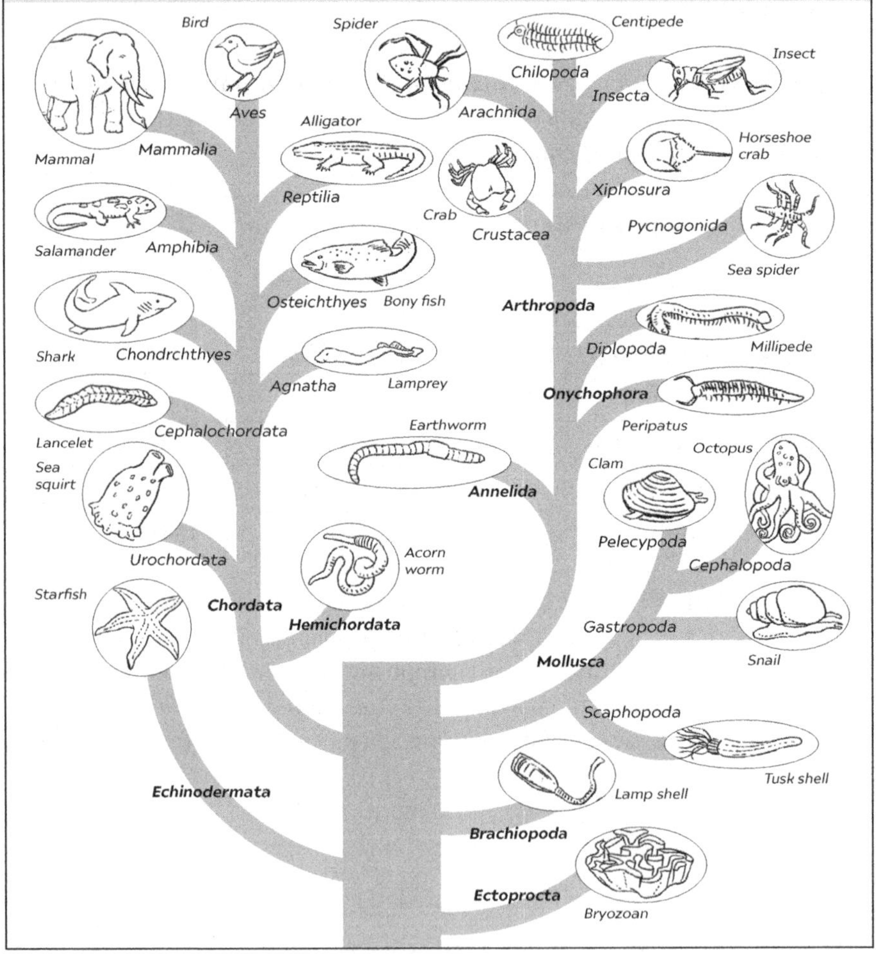

[119]For an in-depth refutation of these failed solutions, see: Stephen C. Meyer, *Darwin's Doubt: The Explosive Origin of Animal Life and the Case for Intelligent Design* (New York: HarperCollins, 2013, 155-289.

vironment, sure, but there is no hard data for macroevolution. Commenting more specifically on molecular evolution, biochemist Michael Behe candidly remarks,

> *"The idea of Darwinian molecular evolution is not based on science. There is no publication in the scientific literature—in journals or books—that describes how molecular evolution of any real, complex, biochemical system either did occur or even might have occurred. There are assertions that such evolution occurred, but absolutely none are supported by pertinent experiments or calculations. Since there is no authority on which to base claims of knowledge, it can truly be said that the assertion of Darwinian molecular evolution is merely bluster."*[120]

So, what does biblical creationism have to offer by way of explaining biological variation? Darwin and Mendel famously showed that certain situations and constraints can draw out information and genetic responses. Leveraging these situations, humans have ingeniously been able to diversify species, methodically selecting desirable adaptations to function in ways that fit the ambitions of the breeder. Remember that this is sometimes termed artificial selection and illustrates the power of intelligent design acting on an already functional, information-laden system. Artificial selection stands in stark contrast to natural selection, which argues that no designer is necessary to draw out these attributes and that information can accumulate by chance and the influence of mutations preserved by natural selection. Biblical creationism uses the well-observed data of artificial selection to highlight the necessity of an informational point of origin at the hands of God. Natural selection extrapolates far beyond the available data to justify an *a priori* anti-theistic and/or anti-biblical stance. Note the following breakdown of the two processes:

To accomplish His goals, ambitions, and purposes, God created functionally complex and disparate organisms that could diversify, within His direct oversight, into further representatives within those initially designated kinds. This is the theological, teleological, cosmological, biological, chemical, and taxonomic takeaway from Genesis 1, especially verses 11-13 and 20-31. The text repeatedly emphasizes the original creation of organisms according to distinctive "kind" and also establishes reproduction according to specific "kind." Countless representations of this are available in extant living things as well as fossils. There are spiders, horses, canines, felines, bears, and numerous other

---

[120] Michael Behe, "Intelligent Design Theory as a Tool for Analyzing Biochemical Systems," in William Dembski, ed., *Mere Creation: Science, Faith, and Intelligent Design* (Downers Grove, IL.: InterVarsity Press, 1998), 183.

Stark Contrasts Between Artificial and Natural Selection		
	Artificial Selection	**Natural Selection**
Goal	Aim (end) in view	No aim (end) in view
Process	Intelligently guided	Blind chance
Choices	Intelligent choice of breeds	No intelligent choice of breeds
Protection	Breeds guarded from destructive processes	Breeds unguarded from destructive processes
Anomalies	Preserves desired anomalies	Eliminates most anomalies
Interruptions	Continued interruptions to reach desired goal	No continued interruptions to reach desired goal
Survival	Preferential survival	Non-preferential survival
This chart has been adapted from: Norman L. Geisler and Frank Turek, *I Don't Have Enough Faith to Be an Atheist* (Wheaton, IL: Crossway, 2004), 143.		

clear-cut animal kind distinctions. Now, what exactly the exact parameters for "kinds" in Genesis 1 may be is not obvious.[121] Some kinds are filled with all sorts of different examples (e.g., felines), while others seem to have only one species (e.g., humans). As discussed earlier, taxonomy is notoriously challenging; the science is tentative, and dogmatism will regularly lead to ridicule as even more discoveries occur.

When we strive to explain biological variations, we have to establish our highest authority, either the fallible field of science or God's infallible Scriptures (Psalm 119:160; Isaiah 40:8; Hebrews 4:12). When we start with the infallible, we can make sense of what we imperfectly observe and often incorrectly interpret (cf. 1 Corinthians 13:12). The portrait of Genesis is of God creating disparate kinds and then directly guiding those kinds into limited change and variation within His appointed parameters. That is why, as advanced as we have become with crossbreeding, we cannot create shark-bears or eagle-elephants (as amusing and simultaneously terrifying as that would be). God has designed intuitive systems able to respond to their changing environments and needs. The presence of these incredibly intuitive biological systems is a breathtaking testimony to the divine Designer (Psalms 19:1; 139:13-17). These

[121]A famous criticism of the historicity of Genesis 1 is Adam's task of naming the creatures on day 6 (cf. Genesis 2:19-20). Critics argue that such a task could never be completed in such a short time window. This criticism improperly assumes that "kinds" refers to each distinct species, conforming also to modern Linnean classification standards. For a brief rebuttal of this criticism, see: Eric Lyons, https://apologeticspress.org/video/was-there-enough-time-on-day-6-of-creation-5888/

observable and testable wonders should inspire all to magnify the Creator of all things for His immeasurable wisdom (Revelation 4:11), not mindless materialistic advancement via evolutionary mutations working in tandem with natural selection.

Concluding Thoughts

Darwinian Theory and many of its newer permutations have failed to sufficiently explain both chemical and biological evolution. When the exponentially increasing time window became prohibitive, evolutionary advocates fled into the shelter of catalyzing factors and formulas that might permit greater changes in shorter amounts of time. The walls of this stronghold have likewise fallen. Such proposed mechanisms prove to have the same origin, efficacy, and economy issues.

No plant, animal, or human has ever been observed to change into another biological form within the upper realms of taxonomy. Every observed and recorded change has been a slight modification in response to environmental demands or to the direct involvement of designing intelligence. Advocates of evolution suggest that these changes act in a combinatory fashion over time to elicit major changes in higher levels of classification. This process is aided, so the story goes, by the unseen hands of mutations working through natural selection. This is an interpretational jump without sufficient scientific warrant and has never been observed in the laboratory.

The Scriptures teach us that God engineered biota that can respond to their environment in certain ways, with certain limitations. To achieve His goals of filling the Earth, He created "kinds" which could produce variety without infringing on His ordained "kind" classification. According to the Scriptures, history is not filled with millions of failed experiments and death resulting from those experiments; rather, history is filled with breathtaking variation within kinds as designed by Almighty God in order to populate the antediluvian Earth and subsequently repopulate the postdiluvian Earth.[122] These changes operated either through God's originally created natural processes or through His direct interventional oversight (Isaiah 42:5; Acts 17:26-28; 1 Corinthians 8:6; John 1:3; Jeremiah 10:12-13).

In our next chapter, we will consider the purposeful designs of some of these kinds and the eschatological ends to which evolution and creation are moving. Let's wrap our current chapter up with a nice summary quote from biochemist Michael Behe:

---

[122]Death entered through Adam (Romans 5:12-21). Theistic evolution teaches that death was here long before Adam.

"The firm conclusion I've drawn over the past decades of research is this: despite occasional questions and bumps along the road, the greater the progress of science, the more deeply into life design can be seen to extend...not only is the Darwinian mechanism devolutionary; it is also self-limiting—that is, it actively prevents evolutionary changes at the biological classification level of family and above...In brief, although chance surely affects superficial aspects of biology, the newest evidence confirms that life is the intended work of a mind and that that work extends much more deeply into life than could previously be seen." [123]

---

[123]Michael J. Behe, *Darwin Devolves: The New Science About DNA That Challenges Evolution* (New York: HarperOne, 2019), 10-11.

Thought/Discussion Questions

1. Give an example of variation across organisms.

2. What are the two primary ways of understanding variation among living things?

3. Is variation proof of evolution?

4. Are mutations always negative?

5. Pick and research one type of genetic mutation. Explain what this type of mutation can and cannot accomplish and whether it is positive or negative in nature.

6. What role does God's creation of "kinds" play in our understanding of observable variation?

7. Is God's guidance and oversight of mutations and variation the same as theistic evolution? Why or why not?

CHAPTER 7

Creation vs. Evolution – Purpose and Destiny

In the general theory of evolution, the heroes are time and chance. Billions of years, mutations, and natural selection collide serendipitously to create life's vibrant tapestry. Animation began from a chance lightning strike in a pool of ooze in the primordial Earth (or, the Earth was seeded by a sample from outer space). It evolved successively from a single-celled organism to a multi-celled organism to plants, animals, and eventually humans. According to this model, humans are merely the "highest" and "most advanced" animal in a long line of evolution. This picture could not be any more different from Scriptural truth.

As we continue to compare the worldviews of creationism and evolution, we turn now to the evidence of design as seen in specified operation, purpose, and function. After considering this, we will then wrap the chapter up with a brief discussion on each view's eschatological outlook.[124]

Contrasting Operation, Purpose, and Functionality

According to Christianity

The biblical portrait of creation is inseparable from its portrait of the Creator. The opening chapters of Genesis briefly detail God's beautiful pattern of creation, endowing each element with orderliness, individuality, and function. Some patterns are more obvious, such as the structurally recurring "golden ratio," which follows the mathematical Fibonacci sequence. Some functions were physically necessary, whereas others were more aesthetic. Other functions are less obvious but, upon more meticulous inspection, reveal illimitable complexity. The Scriptures portray God as a divine tailor providing exact fits, not a slipshod novice haplessly draping lifeless mannequins and trying to figure solutions out along the way (i.e., theistic evolution).

In addition to the prominent tailor-made designs revealed in the opening chapters of Genesis, passages such as Psalm 104 (a passage we highlighted briefly back in Chapter 1) describe calculated functionality. The psalm is so appropriate that we want to quote it here fully:

---

[124]Eschatology is the systematic study (Greek: *logos*) of end times (Greek: *eschaton*).

Bless the Lord, O my soul!

O Lord my God, You are very great:
You are clothed with honor and majesty,
Who cover Yourself with light as with a garment,
Who stretch out the heavens like a curtain.

He lays the beams of His upper chambers in the waters,
Who makes the clouds His chariot,
Who walks on the wings of the wind,
Who makes His angels spirits,
His ministers a flame of fire.

You who laid the foundations of the earth,
So that it should not be moved forever,
You covered it with the deep as with a garment;
The waters stood above the mountains.
At Your rebuke they fled;
At the voice of Your thunder they hastened away.
They went up over the mountains;
They went down into the valleys,
To the place which You founded for them.
You have set a boundary that they may not pass over,
That they may not return to cover the earth.

He sends the springs into the valleys;
They flow among the hills.
They give drink to every beast of the field;
The wild donkeys quench their thirst.
By them the birds of the heavens have their home;
They sing among the branches.
He waters the hills from His upper chambers;
The earth is satisfied with the fruit of Your works.

He causes the grass to grow for the cattle,
And vegetation for the service of man,
That he may bring forth food from the earth,
And wine that makes glad the heart of man,
Oil to make his face shine,
And bread which strengthens man's heart.
The trees of the Lord are full of sap,

The cedars of Lebanon which He planted,
Where the birds make their nests;
The stork has her home in the fir trees.
The high hills are for the wild goats;
The cliffs are a refuge for the rock badgers.

He appointed the moon for seasons;
The sun knows its going down.
You make darkness, and it is night,
In which all the beasts of the forest creep about.
The young lions roar after their prey,
And seek their food from God.
When the sun rises, they gather together
And lie down in their dens.
Man goes out to his work
And to his labor until the evening.

O Lord, how manifold are Your works!
In wisdom You have made them all.
The earth is full of Your possessions—
This great and wide sea,
In which are innumerable teeming things,
Living things both small and great.
There the ships sail about;
There is that Leviathan
Which You have made to play there.

These all wait for You,
That You may give them their food in due season.
What You give them they gather in;
You open Your hand, they are filled with good.
You hide Your face, they are troubled;
You take away their breath, they die and return to their dust.
You send forth Your Spirit, they are created;
And You renew the face of the earth.

May the glory of the Lord endure forever;
May the Lord rejoice in His works.
He looks on the earth, and it trembles;
He touches the hills, and they smoke.

I will sing to the Lord as long as I live;
I will sing praise to my God while I have my being.
May my meditation be sweet to Him;
I will be glad in the Lord.
May sinners be consumed from the earth,
And the wicked be no more.

Bless the Lord, O my soul!
Praise the Lord!

Take a moment to meditate on the complex systems that God has made to work together in specified ways. Think of finely-tuned interplanetary movement and the precise placement of the moon and the sun in exact necessary proximity to Earth within its elliptical orbit (Psalms 8:3; 19:1-6). Think of the micro-cosmic strong and weak nuclear forces that hold atoms together. Think of bio-navigation, particular to different animal kinds such as migratory birds (Jeremiah 8:7). Meditate on the built-in antifreeze systems evident in sundry Arctic and Antarctic species. Add to these the intelligence protocols built into living systems, enabling them to adapt to their environments, which can drastically differ depending on seasons. These innumerable systems are then compounded in difficulty as we consider symbiotic relationships that defy purely natural or even theistic evolutionary explanations. This is especially true regarding obligate symbiosis, in which both organisms require a shared relationship to survive (e.g., algae and fungi in lichen). How could these connections form independently? Who informed them of these necessary biological marriages and hammered out their contractual obligations? How could they survive without their counterparts as they would be required to in evolutionary models?

In fact, biota are so well-designed that humankind strives to model our most ingenious inventions after them through biomimicry.[125] Biological mechanisms for flight found in bumblebees and falcons have inspired countless aircraft architects and engineers (including the Wright brothers and Leonardo da Vinci). Sharks and beavers have inspired the production of diving wetsuits. Termite towers have inspired office building architecture. Burrs have inspired Velcro. Whales have inspired turbines. The list is unending as we have diligently sought to mimic these design systems. And guess what? Every system we

[125]A major part of the argument for God from biomimicry is the efficiency of biological systems. Humans have abjectly failed at achieving these levels of efficiency in their own productions. For more on this, see: J. Warner Wallace, "Why the Efficiency of Biological Organisms Cannot Be Explained by Evolution," *Cold-Case Christianity,* November 18, 2015, https://coldcasechristianity.com/writings/why-the-efficiency-of-biological-organisms-cannot-be-explained-by-evolution/

design turns out to be a cheap imitation of the real thing! The best of human ingenuity is a far inferior version of what we observe in nature!

In one of his newest books, Michael Behe offers a superior synopsis of what we are talking about here. He says,

> *"Unlike that of, say, a field of boulders at the bottom of a mountain, the complexity of living systems demands explanation because they do something, they work, they're functional. For millennia the eye has been a paradigm of teleology in life—something that has an indisputable purpose. Although they have been discovered just recently, the same goes for the other systems described in this chapter—insect gears [referencing a biological system planthoppers use to efficiently coordinate their movement], tank treads [referencing a complex transportation system of certain bacteria], and the like; the purposeful arrangement of their parts is crystal clear. They aren't exceptions; the more research proceeds, the more and more deeply into life teleology can be seen to penetrate."* [126]

Christianity offers a patently clear defense of divinely ordained purpose, especially concerning the superior place of humanity (Genesis 1:26-28). God's designs are truly beyond precise computation. He made the world through His Son (Hebrews 1:1-2), begetting every nation of humanity from one blood (Acts 17:24-31). He has plans for His creation and can and will direct our steps (Proverbs 16:9) into the path of His purposes (Psalm 37:4; Romans 8:18, 28). His key purpose for humanity is not merely physical; He wants people to revere Him and follow His decrees, and He will hold all accountable to this stated purpose (Ecclesiastes 12:13-14; 2 Corinthians 5:10; Hebrews 9:27). While God certainly blessed organisms, and especially humanity, in sexual reproduction, He has given believers the onus of spiritual reproduction (Mark 16:15).

According to Evolution

Specified operation, purpose, and function are all self-evident. Evolutionists rarely challenge these realities. Among others, Ernst Mayr, the famous evolutionary biologist and master of debate, readily acknowledged that nature abounds in processes that lead to an end or goal. He observed, *"The occurrence of goal-directed processes is perhaps the most characteristic feature of the world of*

---

[126]Michael J. Behe, *Darwin Devolves: The New Science About DNA That Challenges Evolution.* (New York: HarperOne, 2019), 63.

living systems." [127] Given this admission, which directly corroborates Scriptures such as Romans 1, how do evolutionists explain such specified complexity?

Since the evolutionary worldview relies heavily on a priori anti-theism, all proposed explanations rely on naturalistic explanations. There is an implicit assumption that life would simply adapt of its own accord to whatever physical and chemical conditions happen to exist. Such assumptions have motivated all attempts to replicate proposed evolutionary adaptations in the laboratory. However, reproduction by means of stacking the deck through unacknowledged intelligent design is a far cry from observable adaptations of the caliber macroevolution requires. Even through the examination of quickly-reproducing fruit flies and bacteria, few positive changes at all have been accomplished in the laboratory, let alone observed in the wild. The basic requisites of chemistry and physics are dismissed as irrelevant background information in place of heavily biased biological hypotheses that largely remain untested and unverified and which are treated as canonized scientific law. The initial conditions of the universe are core assumptions of evolutionary theory rather than something that requires explanation. How convenient.

As discussed in Chapter 4, in the general theory of evolution, the universe sprang essentially out of nothing through random chance. Nature has evolved through eons of time, and it is survival of the fittest. The theory posits a universe devoid of emotion, always cruel and unfeeling in its relentless drive for survival. No inherent meaning exists apart from random biological processes such as mutation and natural selection. Why did this particular organism survive? Because it evolved to its current state with characteristics that made it most likely to survive. The surviving organism is not the "best" or the "most deserving." It is not necessarily the fastest, strongest, or most intelligent. It just happened to have the right qualities and the perfect set of circumstances to endure. Species that have survived are just the result of billions of years of evolution and, as such, can become extinct as higher and more developed life forms are produced. There is no inherent meaning of life or purpose here on Earth except to survive and propagate the species. According to evolution, all appearances of design, specified purpose, and engineered functionality are simply the deceptions of mutations working through natural selection and genetic anomalies.[128] Rather than marvel at design, evolutionists call us to marvel at chance, which has begotten the Goldilocks universe.

Unsatisfied with these problems, many, including evolutionary theorists,

---

[127]Ernst Mayr, "The Idea of Teleology," *Journal of the History of Ideas* 53.1, January-March 1992, 117-135.

[128]For an influential example of an evolutionary treatment of the so-called "delusion of design", see: Richard Dawkins, *The God Delusion* (New York: Houghton Mifflin, 2008).

are beginning to reject the suppression of design intuitions. Stephen Meyer explains:

> *"Since the 1980s a growing number of scientists and scholars have expressed deep reservations about both biological and chemical evolutionary theory, each with their implicit denial of design. And even orthodox evolutionary biologists admit the overwhelming impression of design in modern organisms. To quote Francis Crick again, 'Biologists must constantly keep in mind that what they see was not designed, but rather evolved.'*
>
> *Thus, it seems an acquaintance with biological organisms, to say nothing of the molecular biology of the cell, leads even those who repudiate design to use language that seems incompatible with their own reductionistic and Darwinian perspective—with their official denial of actual design. Although this may ultimately signify nothing, it does at least raise a question. Does the persistence of our perception of design and the use of incorrigibly teleological language, indicate anything about the origin of life or the adequacy of scientific theories that deny (actual) design in the origin of living systems?*
>
> *The persistence of dissenting scientific opinion and the inability of biologists to avoid the language of purpose raises a pardonable curiosity. Have evolutionary biologists discovered the true cause of the appearance of design in living systems, or should we look for another? Should we trust our intuitions about living organisms or accept the standard evolutionary account of biological origins?"*[129]

Contrasting Destiny and Endgame

Life As A Singular Exception to Entropy

The physicist Erwin Schrödinger tersely and humorously defined life as *"something that goes on doing something longer than you would expect it to do."* [130] This is quite obviously true when we consider the counterintuitive continuation of life despite the scientific law of entropy. Entropy is another name

---

[129]Stephen C. Meyer, *Signature in the Cell: DNA and the Evidence for Intelligent Design* (New York: HarperCollins, 2009), 20, 22.

[130]Erwin Schrödinger, *What Is Life? & other scientific essays* (Garden City: Doubleday, 1956). Accessed at: https://archive.org/details/whatislifeothers00schr/

for the second law of thermodynamics, which states that order tends toward disorder, randomness, and confusion. French engineer Sadi Carnot and Austrian physicist Ludwig Boltzmann most notably mapped out this law.

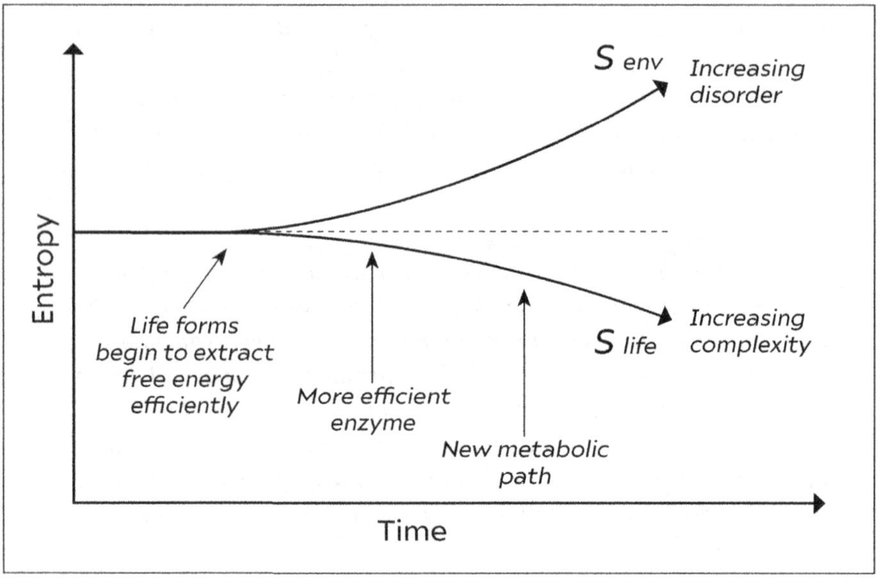

Systemic entropy has been found to be universally true with one lone exception—life! Evolution boldly claims the opposite, specifically, that chaos unassisted moves toward inconceivably complex order without outside regulation![131] This is like a car increasing in fuel the more you drive it. The HMS Beagle runs aground on this scientific law! Contrast biblical creationism, which argues that as sin entered into the world, more and more corruptions began to afflict life on Earth, leading to greater and greater decay. Still, life has navigated these deadly waters and maintained its basic forms over thousands of years. How can this be? The biblical creationist accepts the reality of the God Who holds all things together by the exercise of His benevolent will. In addition, biblical kinds have been able to diversify despite creation's groaning, and humanity has been able to multiply and fill the Earth (Romans 8). Entropy, unchecked, would unhinge the universe; yet, entropy is checked by God, Who commands the Earth to stand fast (Psalm 33:8). This brings a whole new level

[131] *"Evolution argues that biological systems are progressing in an upward direction—progressing from disorganized to highly complex forms of life. Science observes that systems naturally move from order to disorder. The argument for evolution is made against this known and accepted law of science."* (Heath Rogers, *Introduction to Christian Evidences* (Bowling Green, KY: One Stone Press, 2017), 23-24.)

to the archetypal patterning of God as bulwark and shield! (e.g., Genesis 15:1; Deuteronomy 33:29; 2 Samuel 22:3; Psalm 84:11)

The End According to Christianity

Christianity explains the onset of life, the continuation of life, and realities beyond physical death. Since God has designed the world, He is the One Who offers meaning, purpose, and perpetuation. Life is not a hodgepodge of chaos with the illusion of accountability and free will. Life is about God. Life is all about His purpose, His glory, and His goals. From the smallest bacteria moving inside its host to the raccoon in the dumpster to the human snorkeling at the Great Barrier Reef, all were made to exhibit God's grandeur. That is why things continue. That is why chaos does not descend upon us. Because God organized the chaos and fixed His reigns upon the darkness. His Spirit brought order and meaning. For this reason, life is all about praising God! George William Conder penned a memorable hymn explaining this:

All things praise thee, Lord most high,
heaven and earth and sea and sky,
all were for your glory made,
that thy greatness thus displayed,
should all worship bring to thee;
all things praise thee: Lord, may we.

All things praise thee: night to night
sings in silent hymns of light:
all things praise thee: day to day
chants thy power in burning ray;
time and space are praising thee:
all things praise thee: Lord, may we.

All things praise thee, high and low,
rain and dew, and seven-hued bow,
crimson sunset, fleecy cloud,
rippling stream, and tempest loud,
summer, winter – all to thee
glory render: Lord, may we.

All things praise thee, heaven's high shrine
rings with melody divine;
lowly bending at thy feet,

seraph and archangel meet;
this their highest bliss, to be
ever praising: Lord, may we.

All things praise thee, gracious Lord,
great Creator, powerful Word,
omnipresent Spirit, now
at thy feet we humbly bow,
lift our hearts in praise to thee;
all things praise thee: Lord, may we.

Though God created this world and infused it with meaning, purpose, and functionality, He has decreed an end date for this reality. As the psalmist wrote,

"The Lord is merciful and gracious,
Slow to anger, and abounding in mercy.
He will not always strive with us,
Nor will He keep His anger forever.
He has not dealt with us according to our sins,
Nor punished us according to our iniquities"
(Psalm 103:8-10)

Creation is in a state of decay since sin entered the world (Genesis 3). Through the work of Jesus Christ and the Holy Spirit, God has revealed His plan of salvation and other instructions through His Word (Hebrews 1:1-2). These instructions are sufficient to guide us into every good work because they contain all things pertaining to life and godliness (2 Timothy 3:16-17; 2 Peter 1:3-4). Mankind exists to worship the Creator and serve Him, with the prospective destiny of spending eternity with God in heaven. At an unspecified time, Jesus will descend from heaven and destroy Earth in flaming fire (2 Peter 3). He will deal out retribution to those who reject the hope of eternal life and refuse to yield to God (2 Thessalonians 1:7-9). Such persons will be cast out into the outer darkness to suffer in everlasting torment (Revelation 20:11-15). Those who have fallen asleep prior to His arrival will arise and go to the place He has prepared for them to live in eternal glory with the Father (John 14:1-6; 1 Thessalonians 4:13-18). Those alive at His coming will ascend to meet Jesus in the air. If we live in such a way that we are prepared for judgment, through obeying the Gospel and being baptized into Christ (Romans 6:3-5; Titus 3:4-5; Mark 16:16; Acts 22:16; 1 Peter 3:21; etc.), and living a faithful life until death, then we can, by divine grace, spend eternity with God in heaven. The Apostle

John gained a foretaste of this final victory and describes it this way in the last chapter of the Bible:

> *"And he showed me a pure river of water of life, clear as crystal, proceeding from the throne of God and of the Lamb. In the middle of its street, and on either side of the river, was the tree of life, which bore twelve fruits, each tree yielding its fruit every month. The leaves of the tree were for the healing of the nations. And there shall be no more curse, but the throne of God and of the Lamb shall be in it, and His servants shall serve Him. They shall see His face, and His name shall be on their foreheads. There shall be no night there: They need no lamp nor light of the sun, for the Lord God gives them light. And they shall reign forever and ever."* (Revelation 22:1-5)

The End According to Evolution

The eschatological views of evolution are bleak in comparison. You, as a human, may be the current champion, but you are destined to be replaced by another superior product of evolution. You are merely the latest in a long history of imperfect and struggling creatures searching in futility for meaning. Like all things, you will go the way of the earth, returning to the dust. No soul. No spirit. You are only the chance combination of chemicals. You won the evolutionary lottery, but the check bounces because the account has no funds. You will descend into the ground as your ancestors. Your offspring, should you be so lucky, will also descend and arise no more.

In addition to personal nihilism, evolution offers the dreary prospect of "Heat Death," in which the universe expands so that all energy is exhausted, leaving a cold and depleted cosmos. Or, if there is sufficient mass in the universe to eventually halt the expansion, there will be a transition into a compression phase in which a "Big Crunch" will lead to the universe collapsing back in on itself into an incredibly dense point. If chance has its way, perhaps another big bang may occur, and the cycle will start all over. Of course, both of these possibilities may be preceded by the sun finishing its cycle and exploding as a supernova, wiping out all life on Earth, assuming Earth has not already been destroyed by increasing solar radiation and absorption into the sun![132] None of these scenarios fulfills any grand purpose except a purely material universe ending in a purely mechanical way. Such an ending is not the fina-

---

[132]K.-P Schröder, Robert Connon Smith, "Distant future of the Sun and Earth revisited," Monthly *Notices of the Royal Astronomical Society* 386.1, May 1, 2008, 155-163, https://doi.org/10.1111/j.1365-2966.2008.13022.x

	Christianity (Creationism)	Materialism (General Theory of Evolution)
Beginning	In the beginning God created the heavens and the earth (Genesis 1:1)	The "Big Bang," multiple billions of years ago
Earth	The earth was formless and void. God called the dry land earth and the waters that were gathered together He called seas (Genesis 1:2-9)	Over billions of years, the earth coalesced from the disk of matter surrounding the formation of our sun.
Life	The earth (that God had created) brought forth vegetation. God said "let the waters swarm with swarms of living creatures and let birds fly above the earth...let the earth bring forth living creatures according to their kinds"... God created man in His own image. (Genesis 1:11-31)	Life arose unguided out of the primordial ooze on the early Earth's surface, and transformed over billions of years into single-celled organisms, multi-celled organisms, lower and then higher-level animals, and eventually, into humans – solely through mutations and natural selection. The highly information-dense media of DNA and RNA, which contain and carry out the operations of life's blueprint, occurred totally by chance.
Current State	Creation is in a state of decay and decline because of man's fall in the Garden of Eden (Genesis 3). Christians await the return of our Savior, Jesus Christ.	The universe continues to evolve and life continues to evolve. The only thing that matters is survival of the fittest to preserve a specific gene pool.
Future Events	When Jesus returns, each person will be judged according to his/her deeds in the body (Hebrews 9:27-28). Believers in Christ look forward to an eternity in Heaven; the lost are doomed to an eternity in Hell.	When a person dies, they completely cease to exist. Our sun will eventually complete its cycle and explode into a supernova, perhaps in several billion years. The Universe will either expand forever (Heat Death) or collapse in again to restart the cycle (Big Crunch).

le one wishes to share as a motivational speaker or with one's children while tucking them into bed!

What though of the proposed merger between theism and evolution? Perhaps the God Who has remained distant and separated from the universe will decide to get involved in the future? What hope would this offer? Then again, what if He chooses not to get involved? After all, He has been deistically detached from reality to this point. If the creation account needs to be revised and demythologized, perhaps too the biblical portrait of heaven needs to be demythologized? Maybe the "promise of heaven" is merely being remembered for a few years for your sacrifices as a human, destined to be forgotten as the world continues spinning on its axis (or not, who knows?). This, too, offers no consolation or assurance. As the Preacher concluded in Ecclesiastes, life under the sun and without God inevitably reveals a vanity of vanities.

Concluding Thoughts

The best scientific and philosophical explanation for the advent and perpetuation of life is the account of creation recorded in the Bible. The Bible reveals a unique creation, individualized and specialized by order of kinds, within a much smaller time window than evolution requires. The complexity of information we can perceive and measure tells of the Grand Designer, just as lesser marvels of human production can be traced back to human minds.

Why then do many continue to resort to a failed hypothesis? What are they missing? Why can't people submit to what seems so patently obvious? The next two chapters will address these matters and more as we unravel the spiritual components behind this debate.

Thought/Discussion Questions

1. Could God have made the world in a very basic fashion and then walk away from it, allowing for unguided evolution from primordial goo to you? Did He?

2. Define biomimicry and then explain why this is a powerful argument for divine design.

3. What does Psalm 104 indicate about the nature of God's creation?

4. Provide three examples of divine purpose and orderliness.

5. Do evolutionists see evidence of purpose in nature? If so, how would they explain it?

6. Compare and contrast the destinies of evolution and creationism.

7. Does the role of heat at the end of both worldviews mean they are basically the same?

CHAPTER 8

Interpreting Evidence

Science is often naively accepted as pure, objective, and incorruptible. While this is widely accepted in popular conversation and culture, it is entirely untrue. Many factors influence the objectivity of science, especially the interpretation of scientific data. Many experts are aware of and sensitive to these matters, while others seem either ignorant of or else undisturbed by the problems that can and do arise from these dynamics. For those who do recognize these detrimental influences, claims that "Science proves x" are quickly followed by inquiries regarding safeguards and methodology. This is not to impede scientific advancement but rather to keep scientists pursuing objectivity, an ever-present individual and organizational struggle.

In this chapter, we want to discuss what science is and how it requires interpretation. We will then consider factors that influence the interpretation of scientific data. Finally, we will examine how certain interpretations are allowed while others are systematically forbidden. This will set the stage for the next chapter, where we will perform a comparative analysis of the evolutionary worldview and the biblical creationist worldview. Our final chapter will explore the biblical hope of a new creation through the Lord Jesus Christ.

What Is Science and Why Does It Need Interpretation?

Defining science may sound straightforward; in truth, it poses quite a challenge. The layman resource, Dictionary.com, offers a twofold colloquial definition of science: *"a branch of knowledge or study dealing with a body of facts or truths systematically arranged and showing the operation of general laws; systematic knowledge of the physical or material world gained through observation and experimentation."* The same resource proceeds to offer this thought in its more technical section:

> *"The investigation of natural phenomena through observation, theoretical explanation, and experimentation, or the knowledge produced by such investigation. Science makes use of the scientific method, which includes the careful observation of natural phenomena, the formulation of a hypothesis, the conducting of one or more experiments to test the*

hypothesis, and the drawing of a conclusion that confirms or modifies the hypothesis."

Note how several elements of these definitions point to the ability of science to adapt as knowledge increases and more data is collected. Hypotheses may end up right, wrong, or in need of modification based on continued inquiry. Note also the discipline's underlying assumption of arrangement and desire for systematization. Taken together, we see that scientists are fallible, experiments can be misinterpreted, and science depends on patterns.

Now, let's take this information and supplement it with the thoughts of a more academic resource provided by UC Berkeley:

"So what, exactly, is science? Well, science turns out to be difficult to define precisely. (Philosophers have been arguing about it for decades!) The problem is that the term 'science' applies to a remarkably broad set of human endeavors, including developing lasers, analyzing the factors that affect human decision-making, and probing the lifecycles of creatures that live in deep-sea vents...the boundaries of science are not clearly defined..."[133]

This resource proceeds to offer an admittedly non-exhaustive "science checklist" that includes the following criteria:

- Focuses on the natural world;
- Aims to explain the natural world;
- Uses testable ideas;
- Relies on evidence;
- Involves the scientific community;
- Leads to ongoing research;
- Benefits from scientific behavior.

This list has merits. It also reflects philosophical choices. Ultimately, it offers the same fundamental understanding cited above. Popularly and academically, science relies on aims, is open to change with ongoing research, and relies on evidence and testability. Science has "accepted" and "rejected" behaviors, philosophically imposed limits, and yields to the feedback of the scientific community. Often, metaphysics and theology are ostensibly divorced from

[133]"Understanding Science 101: What Is Science? A Science Checklist," *Understanding Science*, UC Museum of Paleontology, https://undsci.berkeley.edu/understanding-science-101/what-is-science/a-science-checklist/

scientific inquiry, but that is untrue based on its methodologies and application. For example, later in Berkeley's educational series, there is a pronounced unwillingness to allow science to lead to any conclusions about metaphysics or theology. This is ironic, given the stated difficulties of establishing clear-cut disciplinary parameters! The only well-established parameter is that science absolutely cannot point to God—isn't that interesting?[134] In logic, that is called "stacking the deck." As Solomon advised, *"He who answers a matter before he hears it, it is folly and shame to him"* (Proverbs 18:13). Perhaps, too, we should consider the statement found earlier in that chapter: *"A fool takes no pleasure in understanding, but only in expressing his opinion"* (18:2, English Standard Version). And again, we might consider 2 Timothy 3:7, which speaks of those who are *"always learning but never coming to the truth."*

We are beginning to see the important role of interpretation and worldview. Scientists are trained to assess, hypothesize, and measure data to provide a well-reasoned explanation. The process of science is to collect valid data with as much detail as possible and to formulate viable theories to explain that data. This data can often be pieced together and interpreted in many ways to form models of how the world works, how the human body works, or how the stars and planets move around in the cosmos. As noted in the definitions above, scientists are looking for patterns and are seeking to understand why those patterns are in place. The big question is: why are the patterns there?

Some scientists look at the wonderful world around us and conclude that the available information points to a Creator (Hebrews 3:4). Other scientists conclude that the world is purely materialistic and mechanical, ruled by chance and serendipity (Romans 1:18-25). In either case, scientists are looking at the same evidence. Trying to resolve this disparity, many believe that scientific inquiry should be provisionally atheistic (Proverbs 14:12). However, why couldn't the resolution be provisional theism? Why does it automatically have to be atheistic? After all, scientists are assuming regularity, structure, and systematization. Does not theism offer those in greater supply than the inescapable chaos that exclusionary naturalism offers?

---

[134]Richard Dawkins, arguably the most famous of atheists, said, *"Biology is the study of complicated things that give the appearance of having been designed for a purpose."* (Richard Dawkins, *The Blind Watchmaker* (New York: W.W. Norton & Company, 1996), 1). Francis Crick, co-discoverer of the double helix structure of DNA, likewise said, *"Biologists must constantly keep in mind that what they see was not designed, but rather evolved."* (Francis Crick, *What Mad Pursuit: A Personal View of Scientific Discovery* (Basic Books, 1990), 138.)

Quotes from Scientists Who Believed in a Creator		
Albert Einstein	**1879-1955**	"I want to know how God created this world, I am not interested in this or that phenomenon, in the spectrum of this or that element. I want to know His thoughts, the rest are details."
Galileo Galilei	**1564-1642**	"The laws of nature are written by the hand of God in the language of mathematics."
Isaac Newton	**1642-1727**	"The most beautiful system of the sun, planets, and comets could only proceed from the counsel and dominion of an intelligent and powerful Being."
James Clerk Maxwell	**1831-1879**	"Almighty God, who has created man in Thine own image, and made him a living soul that he might seek after Thee, and have dominion over Thy creatures, teach us to study the works of Thy hands, that we may subdue the earth to our use, and strengthen the reason for Thy service; so, to receive Thy blessed Word, that we may believe on Him whom Thou has sent, to give us the knowledge of salvation and the remission of sins. All of which we ask in the name of the same Jesus Christ, our Lord."
Johannes Kepler	**1571-1630**	"I had the intention of becoming a theologian... but now I see how God is, by my endeavors, also glorified in astronomy, for 'the heavens declare the glory of God.'"
Louis Pasteur	**1822-1895**	"Science brings men nearer to God."
Max Planck	**1858-1947**	"Religion and science demand for their foundation faith in God. For the former (religion), God stands foremost; for the latter (science), at the end of all thought, For religion He represents a basis; for science, a crowning solution towards a world view."
Michael Faraday	**1791-1867**	"The book of nature which we have to read is written by the finger of God."
Nicholas Copernicus	**1473-1543**	"How exceedingly fine is the godlike work of the Best and Greatest Artist."
Robert Boyle	**1791-1867**	"We ought, whenever we speak of God, and of His attributes, to stand in great awe."
William Thomson Kelvin	**1824-1907**	"As the depth of our insight into the wonderful works of God increases, the stronger are our feelings of awe and veneration in contemplating them and in endeavoring to approach their Author... So will he [the earnest student] by his studies and successive acquirements be led through nature up to nature's God."

Quotes taken from: https://www.christianheritageedinburgh.org.uk/category/science/quotes-from-famous-christians-in-science-and-medicine/

This is all beginning to sound quite philosophical, so let's ground it in something more relatable. Have you ever heard someone accuse a Christian who questions evolution of "not believing in science"? Though science, in theory, feigns theological neutrality, these sorts of responses reveal something quite different about science: the faith component that results from the need to interpret information. After all, who issues that challenge about any other educational discipline? "Don't you believe in literature?" "Don't you believe in health?" "Don't you believe in mathematics?" Honestly, these questions would appear ridiculous if asked. Information leads to interpretation, and interpretation is directed, filtered, and channeled by worldview. Science, then, is often about much more than the core discipline!

Assertions of objectivity in science and theological detachment are often smokescreens for anti-supernatural worldviews.[135] This is ironic, given the history of scientific inquiry and the incredible advances made by those driven by strong theistic persuasions. This is also ironic when you consider modern chaffing against design science. Science looks for organization and signs of intelligent engineering all the time.[136] Science can and does measure intelligence scientifically, yet creation science is often rejected and ridiculed as "pseudoscience."

G.A. Kerkut offers a helpful summation of where this leaves us:

> *"It is in the interpretation and understanding of the factual information and not the factual information itself that the true interest lies. Information must precede interpretation, and it is often difficult to see the factual data in perspective. If one reads an account of the history of biology such as that presented by Nordenskiöld (1920) or Singer (1950) it sometimes appears that our predecessors had a much easier task to discover things than we do today. All that they had to do was realize, say, that oxygen was necessary for respiration, or that bacteria could cause septicemia or that the pancreas was a ductless gland that secreted insulin. The ideas were simple; they just required the thought and the experimental evidence! Let us have no doubt in our minds that in twenty years or so time, we shall look back on many of today's problems and make similar observations. Everything will seem simple and straightforward once it has been explained. Why then cannot we see some of these solutions now? There are many partial answers to this question. One is that often an in-*

---

[135]Perhaps an allusion to 1 Timothy 6:20 in the KJV would be appropriate here, where "science" (i.e., knowledge) can be falsely labeled by the rebellious mind

[136]For example, anthropologists look for signs of intelligence when looking at rocks to see if they were used as tools.

correct idea or fact is accepted and takes the place of the correct one. An incorrect view can in this way successfully displace the correct view for many years and it requires very careful analysis and much experimental data to overthrow an accepted but incorrect theory. Most students become acquainted with many of the current concepts in biology whilst still at school and at an age when most people are, on the whole, uncritical. Then when they come to study the subject in more detail, they have in their minds several half-truths and misconceptions which tend to prevent them from coming to a fresh appraisal of the situation. In addition, with a uniform pattern of education most students tend to have the same sort of educational background and so in conversation and discussion they accept common fallacies and agree on matters based on these fallacies."[137]

Other Factors That Influence Scientific Analysis and Interpretation

We have broadly identified worldview and philosophical biases as determining factors to scientific interpretation. Let's consider other central factors in the interpretation of raw scientific data.[138]

Organizational Sponsorship, Financing, and Tenure

It takes money to live (Ecclesiastes 10:19). Everybody needs to eat and have a place to lay their head. To this end, people do all sorts of things for money, and money does all sorts of things to people! (1 Timothy 6:9-10) This includes scientists. Researchers need money and resources not only to pay their bills and keep food on the table but also to finance their experiments and study interests. This brings with it the temptation to compromise and, often, the obligation to read and report findings in ways that conform to institutional ideologies that will guarantee future financing. There can be major conflicts of interest unless one can self-finance.[139] Obviously, this temptation can occur regardless of one's views on evolution or creationism, so the challenge is not exclusive to evolutionary scientists. Likewise, researchers need organizations

[137]G.A. Kerkut, *Implications of Evolution* in International Series of Monographs on Pure and Applied Biology, volume 4 (New York: Pergamon Press, 1960), 156.

[138]For helpful tips on interpreting scientific claims, see: William J. Sutherland, David Spiegelhalter, Mark Burgman, "Policy: Twenty tips for interpreting scientific claims," *Nature* 503, 335-337 (November 20, 2013).

[139]But even if one can self-finance, that may open the door to projecting one's own biases onto the interpretation of experiments!

and publishing houses to print and market their findings, papers, and books. The risk of publishing on evolution tends to be minimal, as evolution is widely deemed incontrovertible. The more significant challenge is levied against the adversary of evolution, especially one who dares to offer an argument from intentional design.

These added burdens have led many to try and amalgamize the two viewpoints in such a way as to be faithful to their religious and worldview convictions while simultaneously maintaining relations with their research partners and financiers (2 Corinthians 6:14-7:1; Jeremiah 17:9-10). Repeated compromise leads many far afield from where they began, and compromisers find themselves treading a slippery slope. One concession begets another, and then another until a believer can hardly be distinguished from an unbeliever. Psalm 119:1-8 offers insight into the duties of fidelity to God and His Word and to avoiding compromise:

Blessed are the undefiled in the way,
Who walk in the law of the Lord!
Blessed are those who keep His testimonies,
Who seek Him with the whole heart!
They also do no iniquity;
They walk in His ways.
You have commanded us
To keep Your precepts diligently.
Oh, that my ways were directed
To keep Your statutes!
Then I would not be ashamed,
When I look into all Your commandments.
I will praise You with uprightness of heart,
When I learn Your righteous judgments.
I will keep Your statutes;
Oh, do not forsake me utterly!

Hubris and Fear

The self-proposed atheist Bertrand Russell once made an insightful remark, *"The fundamental cause of the trouble is that in the modern world the stupid are cocksure while the intelligent are full of doubt."*[140] His original statement was in reference to the Nazi takeover of Germany and the brainwashing of an entire

---

[140]Bertrand Russell, "The Triumph of Stupidity," in *Mortals and Others* (New York: Routledge, 1998), 28.

population. His words also point to two extremes in which humans are predisposed to making mistakes. Confidence in what is known can be a great blessing, and doubts when one should have confidence can be one's undoing. On the other hand, hubris can cloud judgment, giving the impression that the fool is a genius, while timidity on the part of true geniuses can lead to the dictatorship of fools. Hubris in science is so typical of scientists that "the mad scientist" has become a classic literary and cinematic trope. Scientists can, and often do, wrestle with a god complex. Robert Boyle is a classic example of this. In many ways a devoted believer, Boyle's views on scientists as "nature priests" escalated toward scientific elitism and scientism.[141] The biblical exhortation to humility can serve as a corrective to this god complex (Matthew 18:4; 23:12; Luke 14:11; 18:14; James 4:10).

A related issue that impacts scientific interpretation is fear. Scientists can be afraid of what research may lead to, evolutionists and creationists included. Scientists should strive for objectivity and let the results speak for themselves, but with the influence of worldviews, interpretations can bend and twist results to suit one's ego. While this can cut both ways, this seems to be a prime issue for devotees of the evolutionary religion. We might well consider a similar situation in Jesus' ministry. Following Jesus' resurrection of Lazarus, the religious elite were confronted with their own obvious inferiority. Fears arose as their power began to slip from their fingers. In desperation, they did everything in their power to preserve their place and position of influence (John 11:47-53). In many ways, devout Christians today experience this when they challenge evolutionary science. There is a fear that power, money, and fame will be taken away from those whose ideas will be stripped bare and exposed as folly. Admitting that multiple generations of ideas propping up the evolutionary system are a foundation of moist clay upon which a failed superstructure has been erected is a hard pill to swallow.

Having considered just a few of the variables at play in interpreting scientific research, let's review a few key elements of each worldview that lead evolutionists and creationists to draw different interpretations from the same data.

Evolutionary vs. Creationist Interpretations

Explaining Biological Similarities

Evolutionists look at scientific data and form hypotheses based on exclusively mechanistic explanations. One clear example of this is homology. Ho-

---

[141]For more on this struggle and the rise of scientism, see: Steven Shapin, *The Scientific Revolution* (Chicago: University of Chicago Press, 1998).

mology is the philosophical conclusion that similarity automatically proves common natural descent. So, for instance, when the evolutionist looks at human facial structures and compares them to the facial structures of chimpanzees, several obvious similarities point to their near proximity on the Darwinian tree. This conclusion is reinforced by studies such as the Human Genome Project, which concluded an incredibly high percentage of DNA similarity between primates and humans.[142] For Darwin and his ideological children, evolution rises or falls on these similarities. Darwin explains the importance of homology thus: "*We have seen that the members of the same class, independently of their habits of life, resemble each other in the general plan of organization ... is it not powerfully suggestive of a true relationship of inheritance from a common ancestor?*"[143] Darwin explained this assumption by appealing to the so-called pentadactyl pattern saying, "*What can be more curious than the hand of a man, formed for grasping, the leg of a horse, the paddle of a porpoise, and the wing of a bat should all be constructed on the same pattern, and should include the same bones, in the same relative positions?*"[144]

Pentadactyl Pattern

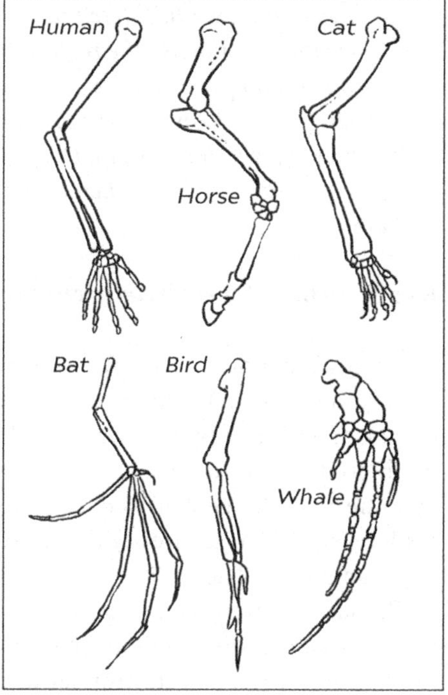

The creationist looks at similar structures but does not conclude the same things. The creationist looks at these configurations and figures that their similarities illustrate derivation from a common Designer. To use an analogy, vehicles from Ford Motor Company exhibit certain observable similarities because they came out of the same factories. In response to evolutionary appeals to the Human Genome Project, creationists appeal to the newer ENCODE Project, which has uncovered much lower

[142]Francis Collins, who oversaw the Human Genome Project, reconciled his theological views with these findings and was content with the theistic evolution explanation of origins and human development despite the challenges discussed in our preceding chapters.

[143]Charles Darwin, *The Origin of Species*, 150th Anniversary Edition (Alachua, FL: Bridge Logos Foundation, 2009), 270.

[144]Ibid.

percentages of similarities between humans and primates and exposed the faulty suppositions of "junk DNA." Creationists conclude that these similarities can be explained by God's created "kinds," which would necessarily utilize templates. Furthermore, creationists appeal to the fact that environments and design goals require particular design plans, which necessarily produce overlap. Dr. Bo Kirkwood summarizes the creationist interpretation of similarity in this way:

> *"So what does unity in type or homology point to if it's not evolution? A common designer is a much more logical explanation for homology than evolution. If God chose DNA to be the underlying blueprint for all life, why should it be surprising that He would use similar homologous structures in His design of animals? For me, in fact, it would be much harder to believe that a common designer would refrain from using similar features in the 'ex nihilo' (creation out of nothing) creation of thousands of varieties of animals. The requirements for life are similar for living organisms and preferred designs are so because they do the function they are designed to do and work! Tires are round because they work well in a car, bicycle, wheelbarrow, motorcycle, etc. Kidneys work because they are suited for their function whether it be in the body of a bear, a horse, or a human."*[145]

Responding to Uniformitarianism

Another major part of evolutionary interpretation is the assumption of uniformitarianism. Evolution advocates ardently hold to consistently uniform processes and thereby hypothesize a long-age Earth that conforms most naturally to a system in clear hostility to biblical creation. The product of this hypothesis is the geologic column to which gradual biological evolution is conformed. We have already discussed some of the core problems of uniformitarianism in Chapter 5.

Biblical creationists appeal to geological evidence of rapid water erosion, massive rapid sedimentation, polystrate fossils, and other data to highlight the significant role of catastrophes in escalating the processes that lead to the geological formations we see today. The most notable catastrophe cited in this interpretation is the global flood of Noah's day discussed in Genesis 6-9. Scientists have also shown that catastrophic processes can rapidly fossilize living creatures in ways that appear to be the result of long-range uniformitarian

---

[145]Dr. Bo Kirkwood, "Creation versus Evolution," *Discovering God's Way* (Athens, AL: Truth Publications, 2017), 19-20.

processes.[146] In addition to this, base strata such as the Cambrian layer exhibit fossils bearing immense complexity, challenging the assumption of gradual evolution.

Coping with Observable Age

Related to uniformitarianism is the observable age of the Earth. Evolutionists explain this apparent age as illustrating successive eras of development beginning from a chance expansion, such as the Big Bang, and leading to fortuitous evolutionary leaps by means of gradual or punctuated processes. In addition to the geological features interpreted by uniformitarianism convictions, there is also the presence of starlight from far distant reaches of space. Given the observed speed of light, the naturalist concludes a timeline of billions of years, a timeline indispensable to evolution. This conclusion has the initial appearance of wisdom but, upon examination, is exposed as faulted (Proverbs 18:17).

The biblical creationist, who only has a timeline of thousands of years given a straightforward reading of the Scriptures, looks to Genesis to resolve the conundrum. Realizing that our insights are imperfect, we entrust the provided explanation of God (Isaiah 55:8-9; Jeremiah 10:23). The proposal of biblical creationists is couched in the idea of apparent age, not actual age. God's design required functioning processes such as immediate starlight and the immediate reproduction of plants and animals. In short, the universe could have been formed in a mature state, even with the existence of expanding galaxies (Hebrews 11:3).

Disallowing Alternative Interpretations

Many factors play into how science is done and how it is interpreted. Because of this, there can be animosity toward certain methodologies and interpretations. This is nowhere more pronounced than with respect to evolutionary science and creation science.

We have mentioned in previous chapters that evolutionary science has spread into every discipline of thought.[147] Despite initial hesitancies to em-

---

[146]For example, see: Alicia Newton, "High pressure fossils," *Nature Geoscience* (October 18, 2007), https://doi.org/10.1038/ngeo.2007.27

[147]Note one author's conclusion: *"There is not a single field of scientific and academic study which has not been greatly modified by the concept of evolution. It provided a new ap*proach to astronomy, geology, philosophy, ethics, religion, and the history of social institutions" (E.G. Bewkes, et al., *Experience, Reason and Faith* (New York: Harper Brothers, 1940), 549).

brace the system, it is now ubiquitously, and often uncritically, accepted. Whether you are confronted by the geologic column on your trip to the Grand Canyon, biological evolution at the San Diego Zoo, or the billions of years in space history at your local planetarium, evolution has its fingers everywhere you go and in everything you see. A little leaven leavens the whole lump (cf. 1 Corinthians 5:6; Galatians 5:9). Books, magazines, journals, and other literature raise the cup to the lips of adults and young children and multitudes drink down the sweet drink, not realizing that it is as dangerous as a drink offered by Jim Jones. To retain this philosophical dominance, personal and organizational advocates of the evolutionary worldview have perpetuated incorrect ideas about creation science, leveraging cultural and even judicial powers to enforce these ideas.[148] Let's consider just a few of these mischaracterizations and vitriol.

"Creation Science and Intelligent Design Are Pseudoscience"

Despite many of creation science's conclusions being deduced scientifically, it is often labeled pseudoscience. One popular example of this is taken from an ACLU website:

> *"Intelligent design (ID) is a pseudoscientific set of beliefs based on the notion that life on earth is so complex that it cannot be explained by the scientific theory of evolution and therefore must have been designed by a supernatural entity...ID is not a scientific theory and therefore cannot be put forward as an alternative to the scientific theory of evolution. ID has no explanatory power or predictive power. It simply says that some things that seem very complex could not have happened based on natural causes. So where it sees complexity, it declares that it must have been created by a supernatural entity. This is not science."*[149]

Wikipedia offers another popular example:

> *"Intelligent design (ID) is a pseudoscientific argument for the existence of God, presented by its proponents as 'an evidence-based scientific theory*

---

[148]Influential court cases addressing the application of creation science in education include: Edwards v. Aguillard, Kitzmiller v. Dover Area School District, and The State of Tennessee v. John Thomas Scopes.

[149]"Frequently Asked Questions About 'Intelligent Design,'" ACLU of Pennsylvania, September 16, 2005, https://www.aclu.org/documents/frequently-asked-questions-about-intelligent-design

about life's origins'. Proponents claim that 'certain features of the universe and of living things are best explained by an intelligent cause, not an undirected process such as natural selection.' ID is a form of creationism that lacks empirical support and offers no testable or tenable hypotheses, and is therefore not science."[150]

More technical dismissals of creation science and ID are seen in resolutions issued by the American Association for the Advancement of Science, which regularly denounce it as pseudoscience on the grounds that it lacks research, is dangerous, and needs to be stamped out by educators. Interestingly, these patronizing statements are often issued unscientifically and do not even necessarily represent the scientists who are a part of the organization![151]

Of course, such characterizations are patently incorrect and fly in the face of combinatory teleological research across centuries.[152] The conclusion of intelligence from design is well established by every observation ever made. William Paley argued this centuries ago, and it is the formative premise that guides much of modern research. Articles, books, interviews, videos, and thousands of other resources are in the mainstream. More is written every day on the subject, appealing to bacterial structures, biomechanical systems, molecular machinery, and a host of other subjects. Hypotheses are regularly formed and tested in ID.[153] Given this reality, these mischaracterizations are patently false, and even though they go down like tasty trifles, they destroy the inner man (Proverbs 18:8; 26:22). As the famous rocket scientist Wernher von Braun commented,

"the vast mysteries of the universe should only confirm our belief in the certainty of its Creator. I find it as difficult to understand a scientist who does not acknowledge the presence of a superior rationality behind the

---

[150]"Intelligent Design," Wikipedia, accessed October 17, 2023, https://en.wikipedia.org/wiki/Intelligent_design

[151]One of the most popular of these resolutions may be accessed at: https://www.aaas.org/news/aaas-board-resolution-intelligent-design-theory

[152]For a brief survey of the history behind the design argument, see: Stephen C. Meyer, "A Scientific History and Philosophical Defense of the Theory of Intelligent Design," *Religion – Staat – Gesellschaft: Journal for the Study of Beliefs and Worldviews, vol. 7,* October 7, 2008, https://www.discovery.org/a/7471/

[153]Stephen C. Meyer, "Yes, Intelligent Design Is Detectable By Science," *Sapientia Journal,* April 26, 2018, accessed at: https://stephencmeyer.org/2018/04/26/yes-intelligent-design-is-detectable-by-science/

universe as it is to comprehend a theologian who would deny the advances of science."[154]

The "God-of-the-Gaps" and Circular Reasoning Challenges

In addition to the popular mislabeling of creation science and ID, some evolutionary advocates and organizations attack creationism with accusations of logical fallacy. One such challenge is that creationism relies on a "god-of-the gaps" fallacy. This fallacy is grounded in the idea that, in the absence of knowledge, one automatically invokes a supernatural agent. While some creation scientists may have engaged in this fallacy, one can just as easily levy the charge at evolutionary scientists. Even Darwin himself posited that the absence of evidence of evolution in the fossil record would be resolved in time. He appealed to an unobservable source in which the solution to his problems could be found. In other words, Darwin fled to his own "god-of-the-gaps." Each accusation and situation then needs to be individually examined without false generalization.

Another logical fallacy that creationists are sometimes accused of is circular reasoning. Some creation scientists may have engaged in this fallacy, but again, many evolutionary scientists have engaged in the same issue! Gary Parker describes a classic example of circular reasoning in Darwinism:

> *"Natural selection is a fact because it's a tautology or truism, a form of circular reasoning. It is argued that the fittest are those that survive in greatest relative numbers and those that survive in the greatest relative numbers are defined as the fittest. That's definitely true, but it's really just an observation, not a profound theory, and begs the question of what makes some organisms fitter than others."*[155]

So what if evolutionists engage in the fallacy? Does this automatically mean that charges against creationists are incorrect? No. However, ID relies not only on complexity but also on specification. Every observable example of this depends on intelligence, and every experiment we have done regarding complexity and specification has yielded the presence of an intelligent agent, Stanley Miller and Harold Urey's famous experiment included. Creationism then relates to universal observation, whereas evolution appeals to unobserved,

---

[154]H.M. Morris, *Men of Science, Men of God: Great Scientists of the Past Who Believed the Bible* (Green Forest, AR: Master Books, 1988), 85.

[155]Gary Parker, *Creation Facts of Life: How Real Science Reveals the Hand of God* (Green Forest, AR: Master Books, 2006), 87.

mindless, and non-complex matter developing minds and complexity. Which then engages in the "god-of-the-gaps" fallacy? Which then is an appeal to circular reasoning? It would seem that it is, in fact, evolution that systematically engages in these logical fallacies.

Active Persecution Against Challengers

The Discovery Institute offers three examples of active organizational and interpersonal persecution against teachers and professors who dare to challenge Darwinian Evolution:

- University of Idaho president Timothy White, issued an edict recently proclaiming that it is now "inappropriate" for anyone to teach "views that differ from evolution" in any "life, earth, and physical science courses."
- Cornell University President Hunter Rawlings, III, delivered a polemic speech denouncing intelligent design and scientists and scholars researching the theory.
- Biology professor P.Z. Myers at the University of Minnesota, wrote this about anyone supporting intelligent design or even just questioning modern evolutionary theory: *"Our only problem is that we aren't martial enough, or vigorous enough, or loud enough, or angry enough. The only appropriate responses should involve some form of righteous fury, much butt-kicking, and the public firing and humiliation of some teachers, many school board members, and vast numbers of sleazy far-right politicians."*[156]

In the same article, they provide three of many illustrations of those suffering at the hands of this "righteous fury." These include:

- Chemistry professor Nancy Bryson lost her job at a state university after she gave a lecture on scientific criticisms of Darwin's theory to a group of honors students.
- Three days before graduate student Bryan Leonard's dissertation defense was to take place, Darwinist professors at Ohio State University accused Leonard of "unethical human-subject experimentation" because he taught students about scientific criticisms of evolutionary theory.

---

[156] "Academic Persecution of Scientists and Scholars Researching Intelligent Design is a Dangerous and Growing Trend," Discovery Institute, December 7, 2005, https://www.discovery.org/a/3077/

- High school teacher Roger DeHart was driven from his public school simply because he wanted his students to learn about both sides of the scientific debate over Darwinian evolution.[157]

Concluding Thoughts

Science can be incredibly objective. Science can also be far from objective. How it is done and interpreted are subjects that many would benefit from studying, whether professionals or non-professionals, Christians or atheists. A better understanding of these areas will contribute to more productive discussions regarding this complex discipline and the complicated phenomena it seeks to understand and explain.

A major part of our discussion in this chapter involves the enduring conflict between two major worldviews. As C.S. Lewis explained years ago in his final published work, our scientific models often are more accurate reflections of the prevalent psychology of our age as much as they are a reflection of actual knowledge. He concluded, *"Nature gives most of her evidence in answer to the questions we ask her."*[158]

---

[157]Ibid. Other famous examples of organizational blacklisting and persecution are related in Ben Stein's 2008 documentary, *Expelled: No Intelligence Allowed.*

[158]C.S. Lewis, *The Discarded Image: An Introduction to Medieval and Renaissance Literature* (Cambridge, UK: Cambridge University Press, 1964).

Thought/Discussion Questions

1. Is science always a purely objective endeavor?

2. Define science.

3. Name the two diametrically-opposed ways that scientists view the world. If you have the chance, watch "Is Genesis History?" which explores these two paradigms.

4. Explain some of the biases that impact scientific interpretation.

5. Select an area where creationist and evolutionist interpretations vary. Explain why each side interprets the data in their own respective way.

6. If evolutionists are so confident in their assertions, why do you think they disallow discussion of alternative interpretations?

7. Have you ever experienced persecution because of creationist beliefs (or any other types of personal beliefs)?

CHAPTER 9

Two Opposing Worldviews

We have mentioned that the general theory of evolution and creationism constitute two opposing worldviews and that theistic evolution results in philosophical contradictions. We want to explore that in more detail in this chapter by defining worldview and examining the main elements of worldview across the divergent paradigms of biblical creationism and evolution.[159]

What Is A Worldview?

Self-revealed by its name, a worldview is the primary means by which we evaluate our world. It is a critical component of our much larger noetic framework, the sum total of everything we think about and know. Many things in our overall noetic framework are of little or no definitive significance (e.g., peanut butter often joins jelly to make a traditional sandwich). Our worldview is a different matter. The fabric of a person's worldview includes only those areas of profound importance. In short, a person's worldview is what that person deems good, wholesome, true, right, and beautiful.

While many models and even definitions of worldview have been advanced, one commonly accepted model contains five vital structural elements: theology, metaphysics, epistemology, ethics, and anthropology. We will use these five worldview pillars to guide our comparison and contrast of the evolutionary worldview with the creation worldview, defining each and then exploring how each worldview views that particular constituent.

[159]For a helpful examination of the paradigms of scientific thought, see: Thomas Purifoy Jr., "The Power of Alternate Histories of the Universe," *Is Genesis History? https://isgenesishistory.com/power-of-alternate-histories-of-universe*

Theology

The foremost part of any worldview is whether there is a deity and what that deity is like. Of course, one may also believe in more than one deity (i.e., polytheism) or even believe that all the various parts of our reality conform to a deity or deities (i.e., pantheism/panentheism). One philosopher offers a great definition of this aspect of worldview:

> *"A world-view will always include either a theology or an atheology. In fact, the most important element of any world-view is what it says or does not say about God. World-views differ greatly over some basic questions. Does God exist? What is the nature of God? Is there but one God? Is God a personal being, that is, is he the kind of being who can know, love, and act? Or is God an impersonal force or power? Because of conflicting views about the nature of God, such systems as Buddhism, Hinduism, Shintoism, and Zoroastrianism are not only different religions; they contain different world-views. Because Christianity, Judaism, and Islam are examples of theism, conservative adherents of these religions hold to world-views that have more in common than they do with dualistic, polytheistic, and pantheistic systems."*[160]

The Biblical Creationist View of Theology

The two forms of revelation provided by God, general and special revelation, guide the biblical creationist view of theology. In Christianity, there is one God Who:

- Created all things (Genesis 1:1; Deuteronomy 6:4; John 1:3; Acts 17:24; Colossians 1:15-19);
- Exists in the unified Godhead (i.e., Trinity) of the Father, Son, and Holy Spirit (Matthew 3:16-17; 28:19; Romans 1:20);
- Is an eternal, immortal, and unchangeable Being of unfathomable power, wisdom, and goodness (Romans 11:33-36; Hebrews 13:8; James 1:17);
- Is utterly unique as the Deity, and all efforts to worship lesser beings, or things, is rejected (Exodus 34:14; Isaiah 45:5);
- Is intimately personal, having revealed Himself in a variety of ways, actively striving for fellowship with His creation and especially humanity (Acts 17:29; Hebrews 1:1-2);

---

[160]Ronald H. Nash, *Faith & Reason* (Grand Rapids: Zondervan, 1988), 30.

- Is spirit and dwells in unapproachable light, yet is not formless (John 4:24; Philippians 2:6; 1 Timothy 6:15-16).

Christianity teaches that despite human limitations and weaknesses, God's desire to be with His people culminated in the incarnation of Jesus (John 1:14; Colossians 2:9) and the gift of the indwelling Holy Spirit to believers (Acts 2:38; 2 Timothy 1:14). The Bible also consistently portrays God as continually interested and involved in creation in both perceivable and imperceivable ways (Acts 17:25-28).

Evolutionary & Theistic Evolutionary View of Theology

While Christianity foundationally requires the existence of a real, personal, involved God, evolution requires no such appeal to the supernatural. That's not to say that some evolutionists do not espouse or even deeply feel committed to belief in God; rather, evolution as a worldview can stand or fall without the conviction that God exists.[161] The evolutionary model bypasses a theological foundation in favor of explaining chemical and biological advancements within the natural order by means of things within the natural order, namely, mutations, natural selection, and other evolutionary contributors. For many evolutionary scientists, methodological naturalism is compulsory to pursuing root causes.[162] Appeals to the supernatural and theology are denigrated as fearful flights to a "god of the gaps" fallacy rather than performing untainted science. For many espousing evolutionists, including most theistic evolutionists, attention to God is quite secondary, to the point that God often becomes inherently deistic in function. God may exist, but He is distant and mostly or totally uninvolved. The position of a creator in the evolutionary worldview is mutation working on natural selection, not the God of the Bible. This theological slant is antithetical to the biblical portrait of God operating and creating in discernible ways. Theistic evolution only compounds this difference by severing theology from the core of history. Theology thus gets shifted to the upper

[161]Richard Dawkins is an avid proponent of atheistic evolution, while Francis Collins is a major proponent of theistic evolution.

[162]Ironically, as evolutionary advocates begin to recognize and vocalize the increasingly tenuous ground of evolution, many leading Christian leaders are pushing for acceptance and adherence to the "increasing evidence for evolution." As the evidence mounts and design and purpose are becoming more evident, intelligent design and creationism should be gaining credence. In philosophical terms, methodological naturalism is losing sway, and theology is coming to the forefront. For a helpful ten-minute synopsis of this movement, as well as a survey of core issues for theistic evolution, see: "The Problem with Theistic Evolution," Crossway, August 9, 2018, https://www.youtube.com/watch?v=GkAxRY41ndU

story where facts and what we can know are replaced by subjective beliefs and feelings.[163] Science and theology are divorced, and worldviews seriously conflict.

Metaphysics

Metaphysics is a term that unfortunately carries a great deal of baggage. For some, metaphysics conjures up strange and curious phenomena, perhaps even the forbidden and taboo. For others, metaphysics brings to mind the complex systems of Plato, Aristotle, Hegel, and other esoteric philosophers. For our discussion, we mean metaphysics in the sense of how a person looks at ultimate reality. Questions in this respect include: How does reality relate to one's theological views? Is the existence of the universe a brute fact? Is the universe eternal? Are there laws to which God must conform in His operations of the universe? Are God and the universe interdependent? Does the universe function only in terms of material, mechanistic means, measures, and methods? Is there function, purpose, and design evident in the universe? Are miracles possible, or is the universe a closed system? Depending on the person, these questions may occupy more or less active focus, but they are all part of the integral structure of one's worldview.

The Biblical Creationist View of Metaphysics

The biblical creationist view of metaphysics is a natural extension of its theology. The central beliefs of Christianity in this respect are as follows:

- In the beginning, God created the universe out of nothing (i.e., *ex nihilo*).
- All of matter and time itself depend on Him (Genesis 1:1).
- By their creation, the universe and all it contains are necessarily subservient to His will.
- The entire world belongs to the Lord, and He is intimately acquainted with all its ways (Deuteronomy 10:14; Psalms 24:1-2; 50:10-12).
- God had the power to create initially, retains the power to grant and sustain life, and continues to exercise the power to extinguish life as He deems appropriate. He is the Author, Creator, and Sustainer of life (Acts 3:15; 17:28; Hebrews 2:10).
- The entire universe is contingent and non-eternal in the biblical para-

[163]Shifting to the upper story is an allusion to Francis Schaeffer's conception of the upper and lower stories of human thought as presented in: Francis A. Schaeffer, *Escape from Reason* (Downers Grove, IL: InterVarsity Press, 2006).

digm (contrast Platonism, which makes the stuff of the universe preexistent).

God's design of the universe includes establishing functional laws by which it can effectively and predictably operate. The regularity and purpose of natural phenomena and the testimony of His historical intervention all point irrefutably to His identity and necessity. However, as Lord, He can supersede those natural laws in ways that bring Him glory and honor. He has done this through direct miraculous involvement (e.g., Matthew 8:14-17; John 9), less-obvious providence (e.g., Esther 1-10), and feeding exceptions to natural law into the system (e.g., 2 Kings 20:8-11). In an effort to woo humanity into a relationship with Him, God has not left Himself without witness (Acts 14:17; Romans 1:18-20). God has in the past freely exercised His exclusive right to destroy this world (e.g., Genesis 6-9). He has foreordained a future destruction of the world by fire (2 Peter 3:10). As such, God calls all people to repent before a final judgment in which His children will be distinguished from Satan's children, resulting in two divergent eternal destinies, heaven and hell (Matthew 25:31-46).

Evolutionary & Theistic Evolutionary View of Metaphysics

Evolutionary metaphysics looks quite different. As a methodologically naturalistic system, evolution begins with the assumption of the preexistence of matter. Materialism, determinism, and gradualism are accepted as immutable.[164] In this worldview, the ultimate realities are those matters of investigation studied by the hard sciences, not God's relationship to the world and His people. For evolution, the world is a vast machine with countless parts and complex processes, which, if we could just study enough, would be scientifically explicable.[165] In this view, while observable phenomena may appear to exhibit design, they are reducible to exclusively natural processes. There is no room for supernatural influences in this closed system. According to the evolutionary worldview, mutations and natural selection are creative and continuous, and provide the essentials for the appearance of design. There is no need to turn to a Designer. As Richard Dawkins so famously pontificated,

---

[164]Advocates of punctuated equilibrium and those who argue for evolutionary catalysts are a prima facie exception to this since even these advocates point to their theorized catalysts as exceptional, viewing gradualism as a rule.

[165]These thoughts are drawn from: William H. Halverson, *A Concise Introduction to Philosophy,* 3rd ed. (New York: Random House, 1976), 394.

biology is the study of complicated things that appear to have been designed for a purpose.

Theistic evolutionists seek to merge the systems by claiming that God can imperceptibly influence the natural order. The theistic evolutionist firmly holds that God is behind the evolutionary process, upholding it in some unspecified way. This is folly, given evolution's claim to a closed system impervious to supernatural influence. JP Moreland summarizes this philosophical contradiction this way:

> *"There are certain understandings of theistic evolution that are logically contradictory. And the most classic one is this: evolution says that the development of life is a completely unguided process that is random and according to natural law, but there's no guiding it in any direction. There's no endpoint that this is moving toward.*
>
> *Well, theistic evolution accepts that. They accept the theory of evolution as an unguided process. But then they say, 'but God somehow guided the process, and that's why we're theistic evolutionists, because we want to hold that God guided the process.' Well, you can't have a process that's guided and unguided at the same time; that is completely contradictory. Now, there are some who will say, 'We believe that the process was in fact guided by God, so it wasn't unguided, but if I'm going to be a good theistic evolutionist, I can't allow anything God did to be detectable, because if we could detect what God did, then theistic evolution would lose to intelligent design and we would have God's nose poking into science.' And so those theistic evolutionists say, 'Well, God guided the process as long as we can never tell what he did.' I wonder if that's any different than saying that he didn't guide it in the first place."*[166]

Epistemology

Epistemology is a fancy word describing what people believe concerning knowledge. One source specifically identifies it as: *"the branch of metaphysics concerned with the nature and limits of human knowledge."*[167] Both basic and complex questions can drive epistemological inquiry: Can we know anything? How do we know anything? Where does knowledge come from? Are our senses reliable? If so, how reliable are they? What is logic, and why do we feel com-

---

[166]"At the Heart of Theistic Evolution, an Inescapable Contradiction," Discovery Science, December 14, 2017, https://www.youtube.com/watch?v=AzR2rnQ9WLs

[167]William Raeper and Linda Edwards, *A Brief Guide to Ideas* (Grand Rapids: Zondervan, 1997), 384.

pelled to observe the rules of logic? How reasonable are our views? Are our reasoning faculties reliable? How do we differentiate states of consciousness? Are intuitions and gut feelings more reliable than sensory perceptions? What is truth? Is truth relative? Is the scientific method the only, or best, way to gain knowledge? Is faith reasonable? Can God reveal Himself to humanity? What is the relationship between God's mind and the human mind? Our answers to these questions form the epistemology component of our worldview.

The Biblical Creationist View of Epistemology[168]

Many biblical texts inform the biblical creationist's epistemological views. Perhaps the principal text is Genesis 1:26-27, which provides the grounds for human preeminence over beasts. Being created in God's image, humans share degrees of knowledge and consciousness unavailable to plants and beasts, transcending purely natural developments. Complementing this, some passages speak to the positive growth of human knowledge in direct proportion to a relationship with God. In contrast, advances gained in rebellion to God ultimately lead to the erosion of the very epistemological infrastructure humans are blessed with. For this reason, God calls humanity to innocent ignorance with reference to carnal thoughts, motives, and ambitions (Romans 16:19; Galatians 5:19-21; Philippians 2:15). Humanity is called to embrace higher and nobler thoughts (Philippians 4:8; Galatians 5:22-23), taking every thought captive for the obedience of faith (2 Corinthians 10:3-5). Such a journey commences with proper reverence and a heart driven by trust in God (Proverbs 1:7; 9:10). Indeed, growth in knowledge is requisite in the journey of discipleship (2 Peter 1:5-11; 3:18), all beginning with the nexus of implanted thinking capacity and the purposeful setting of eternity in the human heart (Ecclesiastes 3:11).[169]

Biblical creationism also has much to say about corruptions that impact human thinking. Sensory issues can affect the perception and evaluation of information. Memory issues and prejudices can filter information, leading to

---

[168]A helpful source for understanding the Christian's epistemology is: Ronald H. Nash, *The Word of God and the Mind of Man* (Grand Rapids: Zondervan, 1982).

[169]This text and others inform John Calvin's concept of *sensus divinitatis*, the idea that mankind naturally and intuitively knows God and longs to be in a relationship with Him. He famously used this to argue against unbelief and most notably explained the idea thus: "*That there exists in the human mind and indeed by natural instinct, some sense of Deity, we hold to be beyond dispute, since God himself, to prevent any man from pretending ignorance, has endued all men with some idea of his Godhead...this is not a doctrine which is first learned at school, but one as to which every man is, from the womb, his own master; one which nature herself allows no individual to forget.*" (*Institutes of the Christian Religion* I.III.I-III). Many reformed epistemologists have followed, with modification, this basic idea (e.g., Alvin Plantinga).

the imperfect interpretation or intentional manipulation of data. Biblical creationism speaks equally against abject skepticism as well as an idealized and defective system of rigid rational foundationalism. Humans have the capacity for thought, but not all thoughts are good, wholesome, or correct. Humans can take in information and acquire knowledge, but they fight an uphill battle against the satanic influence of spiritual ignorance. Christianity holds to the idea that all imperfections of human knowledge are overcome by help from God, Who alone is perfect in understanding. His Spirit reveals His mind in order to adequately equip our spirits to inform our minds (1 Corinthians 2:9-13; 2 Timothy 3:16-17). Francis Schaeffer, the influential philosopher of Christianity, explains this supernatural correction:

> *"[The Bible] presents itself as God's communication of propositional truth, written in verbalized form, to those who are made in God's image. Functioning on the presupposition of the uniformity of natural causes in a closed system, both the secular and the unbiblical theological thinking of today would say that is impossible. But that is precisely what the Bible says it sets forth. We may take, for example, what occurred at Sinai. Moses says to the people, 'You saw; you heard.' What they heard (along with other things) was a verbalized propositional communication from God to man, in a definite, historic space-time situation. It was not some kind of contentless, existential experience, nor an anti-intellectual leap. We find exactly the same kind of communication occurring in the New Testament, as for example when Christ spoke to Paul in Hebrew on the Damascus road. Therefore, on one hand we have the kind of propositional communication God gives in the Scriptures. On the other hand we see to whom this propositional communication is directed."*[170]

Schaeffer continues,

> *"Christianity is a system which is composed of a set of ideas which can be discussed. By 'system' we do not mean a scholastic abstraction, nevertheless we do not shrink from using this word. The Bible does not set out unrelated thoughts. The system it sets forth has a beginning and moves from that beginning in a noncontradictory way. The beginning is the existence of the infinite personal God as Creator of all else. Christianity is not just a vague set of incommunicable experiences, based on a totally unverifiable 'leap in the dark.' Neither conversion (the beginning of the Christian life) nor spirituality (the growth) should be such a leap. Both are firmly*

[170] Schaeffer, *Escape from Reason*, 115-117.

> *related to the God who is there and the knowledge he has given us—and both involve the whole man."*[171]

Evolutionary & Theistic Evolutionary View of Epistemology

Evolution relies on an epistemology predicated on the *a priori* assumption that science is, by far, a superior source of knowledge regarding reality than theology or any other source of information. Methodological naturalism is the outgrowth of this assumption. To attain accurate knowledge, scientists must formulate theories about the world as if it were true that God and the supernatural realm do not exist and that nature acting on its own can produce everything we see. However, if we are really looking for truth, then it's not persuasive to rule out some kinds of causes before you consider all of the evidence and possibilities. Should we take the latest scientific ideas or the textbook orthodoxy of the scientific community to be our ultimate authority on knowledge? Should we give the position of authority to Scripture or the ever-changing tides of the human opinions within the scientific community? (Ephesians 4:14) With evolution, including theistic evolution, there is an unnecessary deference to scientific authority, which necessarily minimizes the Bible as a source of accurate information, belittling the role of God, and specifically the role of the Holy Spirit, in imparting and communicating knowledge. Moreover, evolution regards the development of human knowledge as a purely biological development which may or may not be reliable. All that evolution can guarantee is that human knowledge and conviction are the product of what works, not necessarily that it is correct. This includes the content of human convictions and the physiology of the inner networks resulting in cognition.

Ethics

Ethics is the next worldview component we will consider. It has been the subject of much research and debate across centuries, with writings like Aristotle's *Nicomachean Ethics*, Kant's *Grounding on the Metaphysics of Morals*, and C.S. Lewis's *Mere Christianity*. Ethics is the "why" behind morality. It is the system by which we understand an action to be wrong. Ronald Nash explains this relationship and offers a series of ethics-related questions:

> *"Most people are more aware of the ethical component of their worldview than of their metaphysical and epistemological beliefs. We make moral judgments about the conduct of individuals (ourselves and others) and*

[171]Ibid, 117.

> *about nations. The kinds of ethical beliefs that are important in this context, however, are more basic than moral judgements about single actions. It is one thing to say that some action of a human being like Adolf Hitler or of a nation like Iran is morally wrong. Ethics is more concerned with the question of why that action is wrong. Are there moral laws that govern human conduct? What are they? Are these moral laws the same for all human beings? Is morality totally subjective (like our taste for spinach), or is there an objective dimension to moral laws that means that truth is independent of our preferences and desires? Are the moral laws discovered (in a way more or less similar to the way we discover that 'seven times seven equals forty-nine'), or are they constructed by human beings (in a way more or less similar to what we call human mores)? Is morality relative to individuals or to cultures or to historical periods? Does it make sense to say that the same action may be right for people in one culture or historical epoch and wrong for others? Or does morality transcend cultural, historical, and individual boundaries?"*[172]

The Biblical Creationist View of Ethics

The Bible's ethical basis involves the unification of the other aspects of worldview with the primary focus that God is in a superior position to humanity and, as such, has the right to communicate what is right, good, and proper. The high place of theology in guiding the Christian's standards contrasts humanity's rebellion against his intended function. Biblical terminology denotes the idea of humanity "missing the mark" and "falling short." God's original creation of good people, bearing His imprimatur, has degenerated from this mark by what the Bible calls "sin." Sin is lawless disregard for God's design purpose for humans (1 John 3:4; 5:17). Universally falling short of God's glory (Romans 3:23) has led all people to plummet from God's design for image-bearers (Matthew 5:48). Falling short has led humanity far afield from God's plan of human sanctity (Leviticus 19:2; 20:7; 1 Peter 1:16). Man's departure has also deeply scarred creation, complicating his environment and introducing hostilities that challenge his livelihood (Romans 8). Despite these challenges and the added weight of Satan's temptations to corrupt humanity (Ephesians 6:10-17), God invites all people everywhere to be zealous in doing what is right, good, and proper in order to tap back into the divine image through the mediation of His Son Jesus' sacrifice (cf. Titus 2:11-14; 2 Peter 1:3-4). This ethical plan is explicated by both specific and general moral injunctions in the Scriptures (e.g., Matthew 22:37-40; Romans 12; Ephesians 4-6; 1 Corinthians 13:1-8a; Gala-

---

[172]Nash, *Faith & Reason*, 31-32.

tians 5:22-23; etc.) that are expected of all people across all cultures and times.

Evolutionary & Theistic Evolutionary View of Ethics

While Christianity sees ethics and morality at the core of human creation in the visage of God, evolution has a radically different approach to ethics. Evolution views ethics from the standpoint of population advancement. This is, in effect, why many morally bankrupt human philosophies and totalitarian regimes have leveraged the evolutionary worldview. Eugenicists, in particular, have sought to purge the human gene pool of corruptions and aberrations by maligning "inferior" hosts in order to facilitate "greater" and "purer" humans. For evolution, it's a dog-eat-dog kind of world: kill or be killed, survive or die. This especially becomes problematic when the numbers tip toward majorities who share this might-makes-right ethical philosophy.

Another consequence of evolutionary ethics is the evaporation of personal accountability. When one is the mere product of mindless evolutionary and genetic movements, one can default to the excuse that they are the way they are due to their evolutionary genetics. To what council or throne do they have to answer? And who is there to say that they should be more than their genetics?

Those who oppose such evil applications of evolutionary ethics claim that humans are indebted to the past efforts of humans, which have created a biological make-up predisposed to making advantageous, more obviously approvable choices. In other words, past examples of individual sacrifice and selflessness that benefitted a population at large have profoundly impacted the survivors and beneficiaries of these choices that they have been encoded into the DNA of the population and have conferred an overall survivability advantage. But this begs several questions. For one, how could this impact be reproductively passed on in the case of sacrifice in which the one who actually made the choice ceases to exist? Next, why would the destruction of one life program a separate life to view that activity as beneficiary when the evolutionary goal is specifically to live and not die? This alternate way of looking at evolutionary ethics is actually contradictory to the evolutionary worldview.

Theistic evolutionists seek to borrow from Christianity's ethical ideals and merge them with evolutionary ones. Such an effort strives to unite unapologetic self-preservation at all costs with altruistic self-sacrifice as the living evidence of the Christian walk. Believers cannot abide this pairing. No, Christianity must divorce itself from the evolutionary program and stand alone as the singular explanation of ideal human behavior. C.S. Lewis opined on how biblical ideals inevitably and ubiquitously present themselves in human activities this way:

> *"Think of a country where people were admired for running away in battle, or where a man felt proud of double-crossing all the people who had been kindest to him. You might just as well try to imagine a country where two and two make five. Men have differed as regards what people you ought to be unselfish to—whether it was only your own family, or your fellow countrymen, or everyone. But they have always agreed that you ought not to put yourself first. Selfishness has never been admired."*[173]

Anthropology

"La Scapigliata, trans. The Lady with Disheveled Hair," attributed to Leonardo Da Vinci

The last item for our comparative worldview analysis is anthropology. Quite different from the biological science, anthropology, in this sense, has to do with the nature of human beings. This inquiry into anthropology combines the four other subjects we have discussed and seeks answers to the meaning, value, and destiny of humanity. Anthropology asks questions such as: Do humans have free will, or are they subjected to deterministic forces? Are human beings purely flesh and bone, or are they spiritual? What is the nature of a human soul, if there is one? What is the relationship between the soul and the mind? What is the nature of physical death? Is there life after death? Are heaven and hell real? Is there personal survival after death?

The Biblical Creationist View of Anthropology

To biblical creationists, the essence of humanity is a natural byproduct of the biblical portrait of God. For this reason, biblical creationism intimately intertwines the realms of theology and anthropology. Since humanity is created

---

[173]C.S. Lewis, *Mere Christianity* (New York: Macmillan, 1960), 19. Consider Lewis' similar thoughts on morality in *The Abolition of Man.*

in the image of God (Genesis 1:26-27), it is his appointed and gifted destiny to be like God. This is the formative basis of much that the old and new covenants both focus on in their teaching (e.g., Leviticus 11:44-45; Matthew 5:48; 1 Peter 1:16; et al). God granted humanity free will but, in choosing rebellion, all have universally fallen short of the image of God (Romans 3:23). Left to his wages, man is destined to an eternal separation from the very God Who created him (Romans 6:23). This eternal separation is depicted as hell, an actual, conscious, abiding torment into which the rebel is cast at the final judgment (Matthew 25:41-46).

In God's love, He brought redemption for all who would, of their own free will, submit and bow their knee to His one and only Son, Jesus Christ (John 3:16). The Holy Spirit promises those who choose this one and only way the realization of a heavenly paradise in perfect fellowship and harmony with God (John 14:6). While people are subject to physical forces and laws that God has put into place, they are free to make decisions which bring the consequences of blessings and curses. They also face the daily conflict of the corruption of the flesh intertwined with a sanctified spirit, a war which will only finally be won at death (Romans 7; 1 Corinthians 15; Revelation 21:4). Upon death, those who have remained faithful to God enjoy the best of redemption; contrariwise, those who return to carnality forsake this and will experience an unquenchable fire (Hebrews 10:26-31).

Evolutionary & Theistic Evolutionary View of Anthropology

The evolutionary worldview generally accepts humanity as the pinnacle of life forms; however, this status is not conferred by God but is instead the end product of superior adaptations. This leads to one of the biggest philosophical obstacles between creationists and evolutionists—two very different views of human life. Without the outside attribution of higher purpose and nobler order, humanity is metaphysically of no greater objective value than any other life form. Attempts to view humans otherwise lead to cries of speciesism.[174] In the evolutionary worldview, the preservation of human life is no more important, objectively speaking, than the preservation of any other biological life. A flower, a bush, and a tree all have as much objective value as humans, though many would inconsistently argue for a greater subjective value. Likewise, an

---

[174]Speciesism is a term introduced by the English philosopher Richard Ryder in the 1970s and popularized by the Australian philosopher Peter Singer. The term is used in applied ethics, philosophy, and animal rights to decry the human assumption of superiority to all other species. Christians cannot in good conscience believe that animals and plants are morally equal to humans; however, this does not give the right to humans to mistreat animals. Humans are to be good stewards of creation as per the creation mandate (Genesis 1:28-30; Proverbs 12:10).

amoeba, a titmouse, and a primate all have equal value to a human child. The only real difference when visiting the zoo is that humans are on the other side of the bars. While many attempt to stay consistent with this view, the anthropological claims of Christianity often have a pestering way of creeping in! An advocate of evolution may give theoretical credence to these equalities, but they are utterly unlivable. When given a chance to spare a human life or one of these other life forms, how incredibly insane it would be for anyone to choose against the human life!

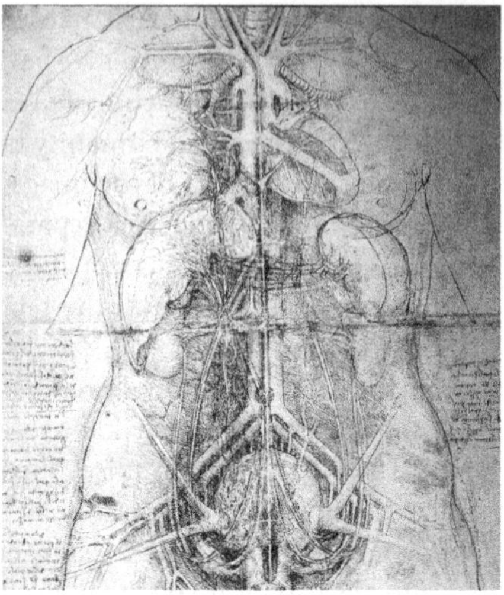
Human anatomy sketch by Leonardo Da Vinci

Concluding Thoughts

Many naively entertain ideas that upend crucial biblical doctrines. This is largely because people do not consider the philosophical underpinnings of systems like the general theory of evolution. Evolution is more than just a one-lane idea restricted to the objective assessment of scientific data. This sort of naivete is evident in theistic evolutionists who do a great disservice to their faith, to the faith of others, and to the sacred revelation of the Holy Spirit when they foolishly embrace the evolutionary worldview.

We want to close with a helpful summary of importance from Ronald Nash:

> *"Whether we know it or not—whether we like it or not—each of us has a world-view. These world-views function as interpretive conceptual schemes to explain why we 'see' the world as we do, why we often think and act as we do. Competing world-views often come into conflict. These clashes may be as innocuous as a simple argument between people or as serious as a war between two nations. It is important, therefore, that we understand the extent to which significant disagreements reflect clashes between competing world-views.*
>
> *World-views are double-edged swords. An inadequate conceptual scheme can, like improper eyeglasses, hinder our efforts to understand the world, and ourselves. The right conceptual scheme can suddenly bring everything into proper focus. But the choice among competing world-views*

Overview of Terms with Sample Questions

Term	Definition	Sample Question
Worldview	The primary lens through which we view the world	What is wholesome, true, right, and beautiful?
Theology	Whether or not god(s) exist(s)	Does God exist?
Metaphysics	How one looks at ultimate reality	Is the universe eternal?
Epistemology	The branch of metaphysics concerned with the nature and limits of human knowledge	Can we know anything?
Ethics	The "why" behind morality	Are there moral laws that govern human conduct?
Anthropology	The nature of human beings	Do humans have free will, or are they subject to deterministic forces?

involves a number of difficult questions. For one thing, we must always contend with the ever-present possibility of nontheoretical factors having an adverse effect on our theoretical judgments. For another, it is difficult to be sure which criteria or tests should be used in selecting among world-views."[175]

[175]Nash, *Faith & Reason*, 33-34.

Thought/Discussion Questions

1. What are the two opposing worldviews discussed in this chapter? Briefly describe each one.

2. What is the foremost element in any worldview? Why is this factor so impactful to every other element of one's worldview?

3. Name three main beliefs in the biblical creationist's view of metaphysics.

4. Briefly describe an evolutionist's view of epistemology.

5. Briefly describe a theistic evolutionist's view of epistemology.

6. How does the biblical creationist's view of ethics compare to the evolutionist's view?

7. What is anthropology? List several questions that an anthropologist might ask.

CHAPTER 10

Becoming a New Creation

There are two dominant worldviews vying for our total allegiance today. One worldview contends that humanity is the product of naturalistic forces that eventually led to his chance development across vast eons of time. The other worldview argues that humanity is specially created in the image of God, given free will, and designed to be holy as God is holy. While some believers attempt to merge these two systems of thought under the guise of theistic evolution, such syncretism is foolish and damnable.

We have offered a brief survey of persuasive evidences for divine design of the universe and a summary of the facts that rule out both the feasibility and the livability of evolution as a worldview. While there is undoubtedly room for adaptation in God's incredibly intuitive designs, evolution cannot offer a well-evidenced basis for origins, nor can it provide an actionable methodology by which the many species of today could have developed. Furthermore, it cannot confer value on your life or on the lives of your loved ones. It is a system inevitably linked to abject and utter nihilism. Who would want that?

It therefore follows that what remains in this final chapter is a final examination of the consequences of biblical creationism, with particular emphasis on personal application. In other words, since we have a Creator, we owe Him our allegiance on His own specified terms as revealed in His exclusive special revelation, the Bible.

The Creator and His Creation

God specially made humanity in His image from the start (Genesis 1:26-27) with the goal that they would use their free will to pursue a relationship with Him voluntarily. A key aspect of this relationship was the extension of His oversight across the world. The late theologian Michael Heiser explained,

> *"Humankind was created as God's image. If we think of imaging as a verb or function, that translation makes sense. We are created to image God, to be his imagers. It is what we are by definition. The image is not an ability we have, but a status. We are God's representatives on earth. To be human is to image God.*

> *This is why Genesis 1:26-27 is followed by what theologians call the 'dominion mandate' in verse 28 ... We are to create more imagers ('be fruitful and multiply ... fill') in order to oversee the earth by stewarding its resources and harnessing them for the benefit of all human imagers ('subdue ... rule over.').*"[176]

As the God and Father of all mankind (Ephesians 4:6), He desired that all would choose Him and enjoy life in Him. For that choice to be real, the possibility of rejecting His love and authority in exchange for a lie had to be an alternative option, so God endowed mankind with free will. He then established indicators in the physical world, perceptible by all, that pointed to His role and humanity's role in relationship to Him (Romans 1:18-20). He added to that general revelation the special revelation of His will for humanity in the Scriptures to lovingly beckon humanity into fellowship with Him. As the famed apologists Norman Geisler and Frank Turek elegantly stated:

> *"One beauty of God's creation is this: if you're not willing to accept Christianity, then you're free to reject it. This freedom to make choices—even the freedom to reject truth—is what makes us moral creatures and enables each of us to choose our ultimate destiny. This really hits at the heart of why we exist at all, and why God might not be as overt in revealing himself to us as some would like. For if the Bible is true, then God has provided each of us with the opportunity to make an eternal choice to either accept him or reject him. And in order to ensure that our choice is truly free, he puts us in an environment that is filled with evidence of his existence, but without his direct presence—a presence so powerful that it could overwhelm our freedom and thus negate our ability to reject him. In other words, God has provided enough evidence in this life to convince anyone willing to believe, yet he has also left some ambiguity so as not to compel the unwilling. In this way, God gives us the opportunity either to love him or reject him without violating our freedom. In fact, the purpose of this life is to make that choice freely and without coercion. For love, by definition, must be freely given. It cannot be coerced. That's why C.S. Lewis wrote, 'the Irresistible and the Indisputable are the two weapons which the very nature of [God's] scheme forbids Him to use. Merely to over-ride a human will (as His felt presence in any but the faintest and*

---

[176]Michael S. Heiser, *The Unseen Realm: Recovering the Supernatural Worldview of the Bible* (Bellingham, WA: Lexham Press, 2015), 42-43. While Heiser took many interpretational leaps that are concerning, His discussion on the image of God in this section of his keystone book is very helpful. This is particularly true with respect to his connections to the philosophical basis for the pro-life movement.

most mitigated degree would certainly do) would be useless. He cannot ravish. He can only woo.'"[177]

Sin Mutates Creation

Sadly, humans have universally chosen rebellion of their own free will (Genesis 8:21; Ecclesiastes 7:20; 9:3). As the Apostle Paul candidly wrote,

> *"For we have previously charged both Jews and Greeks that they are all under sin.*
>
> *As it is written:*
>
> *'There is none righteous, no, not one;*
> *There is none who understands;*
> *There is none who seeks after God.*
> *They have all turned aside;*
> *They have together become unprofitable;*
> *There is none who does good, no, not one.'*
> *'Their throat is an open tomb;*
> *With their tongues they have practiced deceit';*
> *'The poison of asps is under their lips';*
> *'Whose mouth is full of cursing and bitterness.'*
> *'Their feet are swift to shed blood;*
> *Destruction and misery are in their ways;*
> *And the way of peace they have not known.'*
> *'There is no fear of God before their eyes.'*
>
> *Now we know that whatever the law says, it says to those who are under the law, that every mouth may be stopped, and all the world may become guilty before God. Therefore by the deeds of the law no flesh will be justified in His sight, for by the law is the knowledge of sin.*
>
> *But now the righteousness of God apart from the law is revealed, being witnessed by the Law and the Prophets, even the righteousness of God, through faith in Jesus Christ, to all and on all who believe. For there is no*

---

[177]Norman L. Geisler and Frank Turek, *I Don't Have Enough Faith to Be an Atheist* (Wheaton, IL: Crossway Books, 2004), 31. The quote from Lewis is taken from: C.S. Lewis, *The Screwtape Letters* (Westwood, NJ: Barbour, 1961), 46.

difference; for all have sinned and fall short of the glory of God" (Romans 3:9b-23)

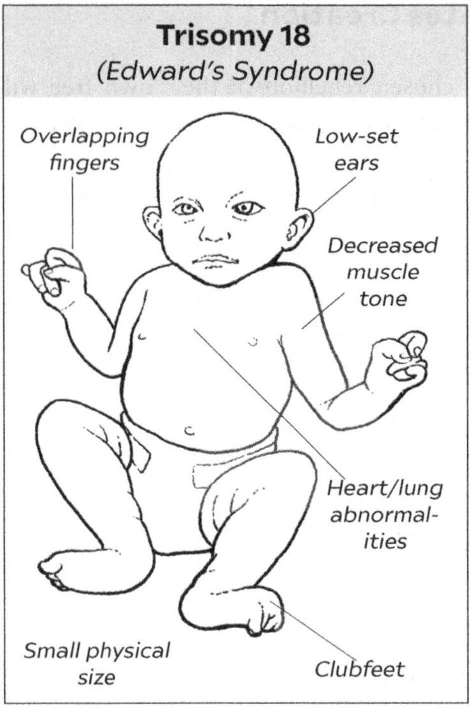

With man's fall from glory came bitter consequences. Death entered into this world, mankind universally stood condemned to eternal torment, and the very world itself became subject to futility (Genesis 3:14-24; Romans 5:12-21; 8:18-25). In many ways, Satan became the ruler of this world (2 Corinthians 4:3-4; John 12:31; 14:30; 16:11; Ephesians 2:2). God's light seemed extinguished. Mankind was cut off from Eden. Yet, there was a glimmer of hope. The God Who brings things into existence from nothing and life from death (Romans 4:17) built hope into the stream of history. Even the universal flood of Noah's day pointed to God's profound ability to transform a thoroughly mutated creation into a new creation (Genesis 6-9). Repeatedly, God began to illustrate that all was not lost. From the new world of Noah's day to the exodus from Egypt (Exodus 1-15), from the resurrection of a man who touched a prophet's bones (2 Kings 13:20-21) to the resuscitation of profligate Israel (Ezekiel 37), God held out a hope of redemption and deliverance. When God stepped into this world in Jesus the Nazarene, He actualized the visible parts of His plan and resurrected His crucified Son to prove to all His power over death and the mutation of sin (Hebrews 2:10-14).

God and the New Creation

The omniscience of God gave Him the clarity to provide a solution to the problem of sin. God is so inconceivably wise that this answer was arrived at before the foundation of the world. Listen to this praise of God's foresight and care for us:

Blessed be the God and Father of our Lord Jesus Christ, who has blessed us with every spiritual blessing in the heavenly places in Christ, just as He chose us in Him before the foundation of the world, that we should be holy and without blame before Him in love, having predestined us to adoption as sons by Jesus Christ to Himself, according to the good pleasure of His will, to the praise of the glory of His grace, by which He made us accepted in the Beloved.

In Him we have redemption through His blood, the forgiveness of sins, according to the riches of His grace which He made to abound toward us in all wisdom and prudence, having made known to us the mystery of His will, according to His good pleasure which He purposed in Himself, that in the dispensation of the fullness of the times He might gather together in one all things in Christ, both which are in heaven and which are on earth—in Him. In Him also we have obtained an inheritance, being predestined according to the purpose of Him who works all things according to the counsel of His will, that we who first trusted in Christ should be to the praise of His glory.

In Him you also trusted, after you heard the word of truth, the gospel of your salvation; in whom also, having believed, you were sealed with the Holy Spirit of promise, who is the guarantee of our inheritance until the redemption of the purchased possession, to the praise of His glory. (Ephesians 1:3-14)

He knew! Before we even decided to abandon His purpose for us, He knew! And He made a way! What unfathomable love and blessing! The greatness of this choice to reclaim us is revealed in the unified desire of all three members of the Godhead to save humanity and their joint efforts to enact the plan of redemption. Note the following truths in this regard:

- The Father brought salvation through His kindness, mercy, benevolence, impartiality, love, wisdom, and power (Titus 3:4-5; Romans 6:23; 1 Timothy 4:10; Ephesians 3:8-19).
- The Son accomplished salvation through His servant leadership, sacrifice, learned obedience, suffering, and compassion (Matthew 1:21; 1 Timothy 1:15; Hebrews 5:8-9; Luke 19:10).
- The Holy Spirit brought salvation through His washing, empowerment, organization, and transformation (1Corinthians 6:11; John 16:7-13; Titus 3:3-5).

This divine collaboration provided a free gift motivated by the joint partnership of God's grace (Acts 15:11; Romans 3:24; Ephesians 2:8-10; Titus 2:11-12), love (John 3:16; 1 John 4:7), and mercy (Titus 3:5). God's desire to save all from sins drove all of this cooperation (1 Timothy 2:4; 2 Peter 3:9).

What does this mean? This means that God set a plan in motion to enable a new creation of His sin-mutated creation, made available in the fullness of times (Galatians 4:4). Through the sacrifice of Jesus and the firstfruits of His resurrection, the gift of God brings about hope that saves (Romans 8:24), faith that saves (Ephesians 2:8), and love for the truth that saves (2 Thessalonians 2:10; 1 Corinthians 13:6). Jesus is the Good Shepherd Who serves as the Door for the sheep to enter in by (John 10). He is the means by which a new birth is possible, redeeming the aborted birth of the carnal man (John 3:5). Through the administration of the life of Christ (Romans 5:10), the cross and the blood of Christ (1 Corinthians 1:18; Galatians 6:14; Matthew 26:28; Romans 5:8-9; Ephesians 1:7, 13-14; Hebrews 9:22; 1 Peter 1:18-19; 1 John 1:7), the Name of Christ (John 20:30-31; Acts 4:10-12; 10:43), and the gospel of our Lord (Romans 1:16-17; 1 Corinthians 15:1-4; James 1:21; 1 Peter 1:23), all people may now choose to enter the way, embrace the truth, and pursue the life made accessible by the inimitable trailblazer Jesus (Acts 4:12). Should they so choose, all people can now become new people in Christ for it is written,

> *"This I say, therefore, and testify in the Lord, that you should no longer walk as the rest of the Gentiles walk, in the futility of their mind, having their understanding darkened, being alienated from the life of God, because of the ignorance that is in them, because of the blindness of their heart; who, being past feeling, have given themselves over to lewdness, to work all uncleanness with greediness.*
>
> *But you have not so learned Christ, if indeed you have heard Him and have been taught by Him, as the truth is in Jesus: that you put off, concerning your former conduct, the old man which grows corrupt according to the deceitful lusts, and be renewed in the spirit of your mind, and that you put on the new man which was created according to God, in true righteousness and holiness."* (Ephesians 4:17-24)

How You Can Become A New Creation In Jesus

Through God's incredible mercy and grace, a new way has opened by which we can return to the original Edenic state where humanity and God are in perfect fellowship.[178] All glory to God for sacrificing His one and only Son so that we are not forever consigned to eternal torment in separation from our Creator! His grace opens the door of salvation that we could not open ourselves (Romans 5:6-8). We earned the wages of death (Romans 6:23), and, unaided, there was no way to earn our return to Eden. No matter how good you or others think you are, no matter what family you belong to, and no matter how often you go to church, none of these things can save you on their particular merits.

We cannot earn our salvation, but God does require that we take part in His redemptive plan so that we can be saved. We must yield to our Creator's design in His appointed way; otherwise, we cannot be saved from our sins. The New Testament tells us that our role, like God's, is beautifully multifaceted. Let's consider several major constituents of our responsibility.

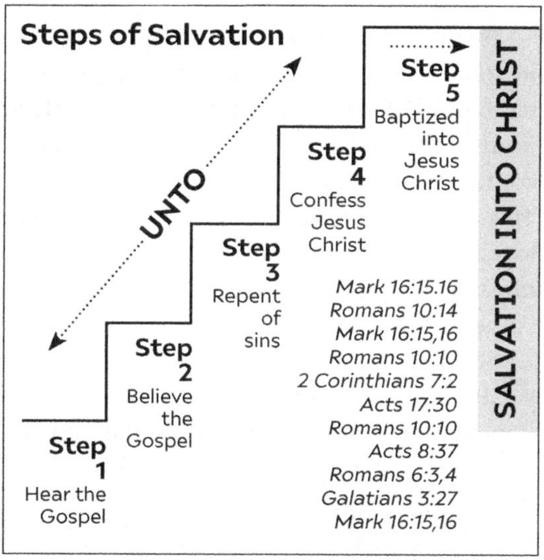

You Must Hear the Gospel and Believe in Jesus Christ

As we take in the Word of God through the sharing of others and through our own meditation on that message, faith grows in our hearts (Romans 10:14-17). The seed of God's Word becomes implanted in our hearts and begins to work on our conscience and spirit (1 Peter 1:22-25). Our reception of that good news opens the door of salvation to us (Acts 11:14). But this is only the start. Many lack *"ears to hear"* (cf. Matthew 11:15), and though they hear the good news, they do not give it an abiding place in their lives. Such a decision is disastrous (Proverbs 28:9; Hebrews 6:4-6). Hearing must produce conviction

---

[178]This beautiful restoration is most notably memorialized by the imagery of Revelation 21-22 which uses the imagery of Eden to depict glorified saints. Other passages such as 2 Peter 3 use similar themes such as the poetic depiction of a *"new heavens and a new earth."*

in the principles of faith revealed in the Scriptures, most notably God's love and sacrifice of Jesus to save His creation (John 3:16). By faith, one will also firmly identify God as Creator and Jesus as the great I AM (Hebrews 11:6; John 8:24). These convictions are integrally bound to the salvation of the soul and they are a key part of our justification before the throne of God (Romans 5:1; Acts 16:31). This belief is essential in salvation, but it is not the only component you are responsible for observing.

You Must Confess the Identity of Jesus and Repent Of Your Sins

If you believe in your heart that Jesus is the Son of God Who came to die in your stead and was resurrected in glory, then you are called to confess this truth unto salvation (Romans 10:9-10; Acts 8:36-37). While this confession of Jesus occurs on the occasion of conversion, there is also a sense in which we confess the identity of Jesus and our relationship to Him in the way we live. This confession leads to a changed life in which our allegiance is evident. This loyalty is so pivotal that even under the threat of persecution and death, we must be willing to confess Jesus, or He will refuse to acknowledge us before the Father (Matthew 10:32-33).

So important is this initial and living confession that it begins to meld into yet another of your responsibilities before God – repentance (Acts 17:30-31). Repentance is not simply fear of the punishment for sin, for Felix possessed that without being saved (Acts 24:25). It is also not merely sorrow (2 Corinthians 7:10). Likewise, prayer, confession of sins, and doing good deeds are not of themselves sufficient to fully accomplish repentance (Proverbs 28:9; 1 John 1:9; Matthew 3:8). Repentance is about subscribing to and developing a godly mindset in which sin is seen as utterly abhorrent and devotion to God is prioritized over the lust of the eyes, the lust of the flesh, and the boastful pride of life (Amos 5:15; Romans 12:9; 1 John 2:15-17). When one repents, they are vowing themselves as a living sacrifice unto the living God (Romans 12:1-2). Repentance blots out sin and brings seasons of refreshing from the presence of the Lord! (Acts 3:19)

You Must Be Immersed in Water and Live Faithfully Until Death

God calls humanity to listen to His message of salvation, to believe in Jesus, to confess His name, and to repent of sins. All of these are inseparable from becoming a Christian. There is still more. God's scheme of redemption is so kaleidoscopic, yet attainable, that it evokes total awe. One of the simplest yet most profound illustrations of this is the culmination of faith in submission to immersion in water to be saved from one's sins. Note the following biblical truths regarding baptism:

- Just as Jesus was laid low in the grave and rose in victory, so too may believers submit to immersion in a watery grave and rise in newness of life (Romans 6:3-6).
- Water baptism is essential to salvation (1 Peter 3:21; Mark 16:16; Acts 2:38).
- Immersion is a circumcision of our spirit as a work of God (Colossians 2:11-12), accomplished by the Holy Spirit who places us in the body of Christ (1 Corinthians 12:13; Titus 3:3-5), and which singularly unifies believers as adopted brothers and sisters of Christ (Ephesians 1:5; 4:4-6; Romans 8:15, 23; Galatians 4:5).
- Baptism is a preparatory purification of the bride for her joining to her Husband (Ephesians 5:26), a calling on God for salvation (Acts 22:16), and an opportunity to be clothed with Christ (Galatians 3:26-27).

Cleansed from our sins, we are called to sanctified living by the grace of God (Titus 2:11-14). This holy living leads to the rejection of sin and corruption, injustice and perversion, and prejudice and partiality (Galatians 5:19-21). It also calls for true religion, sacrificing for the broken and downcast (James 1:26-27), being a voice for the voiceless (Proverbs 31:8-9), and living as royal priests (1 Peter 2:4-10). In this way, we become lights that cast out the luminescence of God into a broken and dying world destined for destruction. Many will glorify God as they witness such things (Matthew 5:14-16). Such a life is as though one no longer even lives, but Christ lives in them (Galatians 2:20; 5:22-23), a wholly new creation (2 Corinthians 5:17). Working out our salvation in daily living is the aim for all Christians, a life lived in full fidelity until one's dying breath (Philippians 2:12; Revelation 2:10).

Concluding Thoughts

God wants to save you, not because you are the fortunate candidate of a divine experiment with evolutionary mutations based on natural selection, but because you are His prodigal son or daughter. He calls you to wholistic love and devotion, untainted by the captivating philosophies of men. Evolution and the proposals of theistic evolution are folly and rubbish. God provides meaning, fulfillment, and purpose in His special creation and in His new creation. He has a destiny for you that involves your glorification in eternal paradise with His Son!

After reading this chapter, you know how to respond. Though you will never be the sole effectual cause of your salvation, you surely must act for yourself. You must recognize your violation of God's will in your life. You alone are accountable and responsible for your choices. You cannot hide in the shadow of someone else and sneak into the back door of Heaven. There are no back doors; there is only one front door into which God freely and kindly invites you. Stop letting others think for you or hold you back. God beckons you. He invites you. Choose life. Choose peace. Choose joy. Choose an eternity of constant satisfaction and deliverance. What other choice could you reasonably make? (2 Corinthians 5:20; James 4:8-10)

As this book ends and you consider your choice, we want to end with the famous lyrics of Carl Boberg's inspiring hymn:

Oh Lord, my God
When I, in awesome wonder
Consider all the worlds Thy hands have made
I see the stars, I hear the rolling thunder
Thy power throughout the universe displayed

And when I think that God, His Son not sparing
Sent Him to die, I scarce can take it in
That on the cross, my burden gladly bearing
He bled and died to take away my sin

When Christ shall come, with shout of acclamation
And take me home, what joy shall fill my heart
Then I shall bow, in humble adoration
And then proclaim, my God, how great Thou art

Then sings my soul, my Savior God to Thee
How great Thou art, how great Thou art
Then sings my soul, my Savior God to Thee
How great Thou art, how great Thou art
How great Thou art, how great Thou art

Thought/Discussion Questions

1. Which of the two dominant worldviews have you been holding closely to, before this study?

2. According to Genesis 1:26-27, what is humankind's special relationship with the Creator?

3. In what ways has sin impacted Creation?

4. What type of future do we look forward to as God's people?

5. What is the first step in becoming a new creation in Jesus?

6. After being immersed and placed in the body of Christ (1 Corinthians 12:13), what sort of life must we endeavor to live?

7. Has this study led you to re-examine your worldview? Has it produced a change in how you plan to live your life from this day forward?

Made in the USA
Monee, IL
04 November 2026

67385040R00095